ADVANCED LOW-POWER

DIGITAL CIRCUIT TECHNIQUES

THE KLUWER INTERNATIONAL SERIES IN ENGINEERING AND COMPUTER SCIENCE

VLSI, COMPUTER ARCHITECTURE AND DIGITAL SIGNAL PROCESSING
Consulting Editor
Jonathan Allen

Other books in the series:

HARDWARE-SOFTWARE CO-DESIGN OF EMBEDDED SYSTEMS: The POLIS Approach
F. Balarin, M. Chiodo, P. Giusto, H. Hsieh, A. Jurecska, L. Lavagno,
C. Passerone, A. Sangiovanni-Vincentelli, E. Sentovich, K. Suzuki,
B. Tabbara
ISBN: 0-7923-9936-6
COMPUTER-AIDED DESIGN TECHNIQUES FOR LOW POWER SEQUENTIAL LOGIC CIRCUITS, J.Monteiro, S. Devadas
ISBN: 0-7923-9829-7
APPLICATION SPECIFIC PROCESSORS
E.E. Swartzlander, Jr.
ISBN: 0-7923-9729
QUICK-TURNAROUND ASIC DESIGN IN VHDL: Core-Based Behavioral Synthesis
M.S. Romdhane, V.K. Madisetti, J.W. Hines
ISBN: 0-7923-9744-4
ADVANCED CONCEPTS IN ADAPTIVE SIGNAL PROCESSING
W. Kenneth Jenkins, Andrew W. Hull, Jeffrey C. Strait
ISBN: 0-7923-9740-1
SOFTWARE SYNTHESIS FROM DATAFLOW GRAPHS
Shuvra S. Bhattacharyya, Praveen K. Murthy, Edward A. Lee
ISBN: 0-7923-9722-3
AUTOMATIC SPEECH AND SPEAKER RECOGNITION: Advanced Topics,
Chin-Hui Lee, Kuldip K. Paliwal
ISBN: 0-7923-9706-1
BINARY DECISION DIAGRAMS AND APPLICATIONS FOR VLSI CAD, Shin-ichi Minato
ISBN: 0-7923-9652-9
ROBUSTNESS IN AUTOMATIC SPEECH RECOGNITION, Jean-Claude Junqua, Jean-Paul Haton
ISBN: 0-7923-9646-4
HIGH-PERFORMANCE DIGITAL VLSI CIRCUIT DESIGN, Richard X. Gu, Khaled M. Sharaf, Mohamed I. Elmasry
ISBN: 0-7923-9641-3
LOW POWER DESIGN METHODOLOGIES, Jan M. Rabaey, Massoud Pedram
ISBN: 0-7923-9630-8
MODERN METHODS OF SPEECH PROCESSING, Ravi P. Ramachandran
ISBN: 0-7923-9607-3

ADVANCED LOW-POWER DIGITAL CIRCUIT TECHNIQUES

by

Muhammad S. Elrabaa
Intel Corporation

Issam S. Abu-Khater
Intel Corporation

Mohamed I. Elmasry
University of Waterloo

KLUWER ACADEMIC PUBLISHERS
Boston / Dordrecht / London

Distributors for North America:
Kluwer Academic Publishers
101 Philip Drive
Assinippi Park
Norwell, Massachusetts 02061 USA

Distributors for all other countries:
Kluwer Academic Publishers Group
Distribution Centre
Post Office Box 322
3300 AH Dordrecht, THE NETHERLANDS

Library of Congress Cataloging-in-Publication Data

A C.I.P. Catalogue record for this book is available
from the Library of Congress.

Copyright © 1997 by Kluwer Academic Publishers

All rights reserved. No part of this publication may be reproduced, stored in a retrieval system or transmitted in any form or by any means, mechanical, photo-copying, recording, or otherwise, without the prior written permission of the publisher, Kluwer Academic Publishers, 101 Philip Drive, Assinippi Park, Norwell, Massachusetts 02061

Printed on acid-free paper.

Printed in the United States of America

To our families

CONTENTS

LIST OF FIGURES		xi
LIST OF TABLES		xix
PREFACE		xxi
1	**LOW-POWER VLSI DESIGN**	1
	1.1 Power Estimation and Evaluation for Digital VLSI Circuits	1
	1.2 Low-Power Impact on Process and Technology	2
	1.3 This Book	3
REFERENCES		5
2	**LOW-POWER HIGH-PERFORMANCE ADDERS**	7
	2.1 Introduction	7
	2.2 Architecture	8
	2.3 Circuit Design and Implementation	11
	2.3.1 TG Implementation	11
	2.3.2 CPL-Like Implementation	11
	2.4 Optimization Using Variable Block Size Combination	13
	2.5 Simulation Strategy	15
	2.5.1 Transistor Sizing	15
	2.5.2 Power Estimation	15
	2.6 Circuit performance	16
	2.7 Layout Strategy	23
	2.8 Experimental Results	23

	2.9	Summary	27

REFERENCES 29

3 LOW-POWER HIGH-PERFORMANCE MULTIPLIERS 31
  3.1 Introduction 31
  3.2 Review of Parallel Multipliers 32
    3.2.1 Braun Multiplier 33
    3.2.2 Baugh-Wooley Multiplier 33
    3.2.3 The Modified Booth Multiplier 34
    3.2.4 Wallace Tree 39
    3.2.5 Multiplier's Comparison 42
  3.3 Multiplier Architecture and Simulation Method 43
    3.3.1 Architecture 43
    3.3.2 Simulation Strategy 45
  3.4 Multiplier Cell 47
    3.4.1 Full and Half Adder Circuits 47
    3.4.2 Multiplexer 58
    3.4.3 The Multiplier Cell Performance 62
  3.5 Booth Encoder 64
    3.5.1 Logic 64
    3.5.2 Circuits 64
  3.6 Add Cell 70
  3.7 CSA 70
  3.8 6-Bit Multiplier 71
    3.8.1 Layout of 6-bit multiplier 73
  3.9 Summary 74

REFERENCES 81

4 LOW-POWER REGISTER FILE 83
  4.1 Introduction 83
  4.2 Architecture and Simulation Procedure 84
  4.3 Memory Cell Circuit 86
  4.4 Write Circuitry 86

	4.5	Read Circuitry	89
	4.6	Decoder Circuit	89
	4.7	32x32-bit Register File	91
	4.8	Summary	96

REFERENCES   97

5   LOW-POWER EMBEDDED BICMOS/ECL SRAMS   99

	5.1	Introduction	99
	5.2	16 Mb$^+$ SRAMs Front-end Optimization	101
	5.3	The Novel W-ORing and Level-Translation Circuits	106
	5.4	The Novel Self-Resetting WL Decoder and Driver	109
		5.4.1 Circuit Operation	112
		5.4.2 Performance Comparisons	112
	5.5	The Novel Latched Sense-Amplifier	113
		5.5.1 Immunity to Bit-line Glitches	118
		5.5.2 Performance Comparisons	119
	5.6	Chapter Summary	120

REFERENCES   123

6   BICMOS ON-CHIP DRIVERS   125

	6.1	Introduction	125
	6.2	The Novel Full-Swing BiCMOS Circuit Technique	127
		6.2.1 Concept of Operation	128
		6.2.2 The Verification of Operation	132
	6.3	Performance Comparisons	137
		6.3.1 Simple Buffers	137
		6.3.2 AND Gates	142
		6.3.3 Master-Slave D Flip-Flops	144
	6.4	Design of the Feedback Circuitry	148
	6.5	Chapter Summary	149

REFERENCES   151

7 INTER-CHIP LOW-VOLTAGE-SWING TRANSCEIVERS — 153
7.1 Introduction — 153
7.2 Low-Power ECL/CML Dynamic Circuit Techniques — 154
    7.2.1 The DAPD Level Shifter — 157
    7.2.2 The DCCP Current Source — 160
7.3 The Universal Transceiver (Receiver/Driver) — 163
    7.3.1 The Universal Receiver — 167
    7.3.2 The Universal Output Drivers — 175
7.4 Chapter Summary — 192

REFERENCES — 193

INDEX — 195

LIST OF FIGURES

Chapter 1

Chapter 2

2.1 Conditional Sum with Carry Select Adder: (a) 4-bit block architecture, (b) implementation of conditional circuit (c) multiplexer TG implementation. 9

2.2 Critical delay path of 4-bit CSA adder. 10

2.3 CSA-CPL-like schematic of the output stage. 12

2.4 Block size effect on performance of 16-bit Adder: (a) equal block size (b) variable block sizes. 14

2.5 Critical carry delay path of: (a) conventional-CLA, (b) CPL-CLA, (c) DPL-CLA (d) TG-CLA, (e) TG-Carry Select (CS), (f) and TG-manchester 17

2.6 Energy versus delay for the minimum transistor size adder (V_{DD}=3.3 V). 18

2.7 Energy versus delay for the optimized transistor size adder (V_{DD}=3.3 V). 20

2.8 Effect of V_{DD} scaling on the adders' delay. 21

2.9 Effect of V_{DD} scaling on the adders' power dissipation at 20 MHz. 22

2.10 Layout illustration of a 4-bit CSA-CPL adder in 0.8 μm technology. 23

2.11 Photomicrograph of (a) Test Chip and (b) 32-bit CSA-CPL adder. 24

2.12 CSA-CLA functionality test; input waveform B_1 (bottom) and output S_{32} (top). 25

2.13 50% time delay measurement for the CSA-CPL adder between B1 and inverted S32 (a) rise time (b) fall time. 26

Chapter 3

3.1	Discrete Cosine Transform	32
3.2	(a) Partial products of a 4 × 4 unsigned integer multiplication; (b) multiplier array; (c) full-adder schematic.	34
3.3	(a) 4 × 4 Baugh-Wooley two's complement regular array (FA : Full-Adder).	35
3.4	8-bit Booth multiplication example using two complement's number.	37
3.5	The previous example of Figure 3.4 with simplified sign extensions.	39
3.6	Block diagram of the $n \times n$ multiplier using a modified Booth algorithm.	40
3.7	Construction of Wallace's tree for an 8 × 8 multiplier: (a) reduction of the 8 partial products with 4-2 compressors; (b) the architecture.	41
3.8	The architecture of 32 × 32 modified Booth multiplier with Wallace tree.	42
3.9	16x16 bit Booth multiplier	44
3.10	Simplified circuit	46
3.11	Multiplier cell (PP-FA)	48
3.12	Schematic of FA circuits	50
3.13	Full adders delay vs power; (a) minimum size (b) optimized for speed	53
3.14	Photomicrograph of FA circuits: (a) conventional, (b) CPL-TG, (c) LCPL2 and (d) Test chip	55
3.15	FAs experimental verification: input waveform (bottom) and output superimposed waveforms (top).	57
3.16	Multiplexer circuits: (a) Single rail (b) Double rail	59
3.17	Delay vs. Power for single rail mux at various output buffer $\frac{W_p}{W_n}$ sizes.	60
3.18	Delay vs. Power comparison of multiplier cells	61
3.19	Comparison of the delay-power product of multiplier cells	62
3.20	Layout of the CPL-TG multiplier cell	63
3.21	Logic and static CMOS implementation of Booth encoder.	65
3.22	CPL implementation of Booth encoder.	66
3.23	Transient response of Booth encoder with and without output buffers	67

List of Figures xiii

3.24 A Power and performance comparison of Booth encoder circuits.   68
3.25 Layout of CPL Booth encoder circuit.   69
3.26 Add cell   70
3.27 6x6 bit Booth multiplier   72
3.28 Effect of FA style on the multiplier performance; (a) delay and (b) power dissipation   75
3.29 Effect of the encoder style on the multiplier performance; (a) delay and (b) power dissipation   76
3.30 Layout strategy   77
3.31 Layout of 6x6 multiplier.   78
3.32 Test chip of the multiplier.   79

Chapter 4

4.1 Register file architecture.   84
4.2 Simplified Schematic for simulation of register file.   85
4.3 Memory cell; (a) schematic, (b) simulation results and (c) layout.   87
4.4 Write circuitry; (a) schematic, (b) clock skew reduction (c) simulation results and (d) layout of the D-FF.   88
4.5 Read circuitry; (a) schematic, (b) simulation results and (c) layout.   90
4.6 AND/NAND circuits; (a) conventional CMOS, (b) HCPL and (c) CPL.   92
4.7 (a) Simulation results, (b) Layout of HCPL circuit.   93
4.8 Register file waveforms; (a) write operation, (b) read operation.   94
4.9 Test chip of the 32x32 bit register file.   95

Chapter 5

5.1 Access time and power of the different generations of CMOS/BiCMOS SRAMs.   101
5.2 The ECL address input buffer and W-OR pre-decoder.   102
5.3 Power vs. CL @ 100 MHz for the ECL and BiNMOS buffers for a fixed delay of 450 pS.   103
5.4 Power versus delay for the ECL and BiNMOS buffers for two values of CL @ 100 MHz.   104

5.5 Speed-up of ECL over CMOS versus the power ratio for several loads. The vertical bars represent the maximum power limit for the ECL buffer. 105

5.6 Speed-up of ECL over BiNMOS versus the power ratio for several loads. The vertical bars represent the maximum power limit for the ECL buffer. 106

5.7 The three W-OR pre-decoder and level-translator combinations. 108

5.8 The total power versus the power of the W-OR pre-decoder for the three combinations of Figure 5.7. 110

5.9 The new word-line decoder and driver (WLDD) circuit. The equivalent logic circuit is also shown. 111

5.10 The output waveform of the new WLDD for different speeds of feedback. 113

5.11 The rise delays of the conventional WLDD circuit (CMOS NAND + BiNMOS driver) and the new WLDD circuit versus the load capacitance. 114

5.12 The novel latched-ECL sense-amplifier. 115

5.13 The simulated performance of the new column sensing technique. 117

5.14 The response of the latched sense-amplifier to two 50 mV bit-line glitches of 100 pS and 200 pS durations. The response to a 75 mV and 500 pS bit-line read-out signal is also shown. 119

5.15 A conventional ECL sense-amplifier with cross-coupled PMOS loads. 120

5.16 The delay of the two sense-amplifiers versus their average power. 121

Chapter 6

6.1 Some of the existing BiCMOS circuits. 126

6.2 The conventional BiCMOS pull-up circuit; (a) schematics, (b) the transient response of the base and output voltages, and (c) the transient response of the base and emitter currents. 129

6.3 Different implementations of the novel BiCMOS pull-up circuits utilizing the positive dynamic feedback. 131

6.4 A cross-section of the merged BiPMOS device used in the transient device simulations. 132

List of Figures

6.5 The pull-up transient response of the novel circuit for, several technologies and supply voltages (solid lines), and for $C_L = 0.3pF$. For the (0.8 μm, 5V) and (0.5 μm, 3.3V) technologies, the dotted lines represent the response of the circuit in [5] with an MOS output shunt.   133

6.6 The emitter current of the pull-up BJT during pull-up and pull-down at a 250 MHz frequency.   135

6.7 The transient response of the novel pull-up circuit with and without turning the PMOS off. Also the effect of C_{fb} is shown.   136

6.8 The two parasitic PNP BJTs in the merged BiPMOS structure.   136

6.9 The average delay of the circuits in Fig. 6.3 compared to the optimized CMOS buffers and using the $(0.2\mu m, 2V)$ technology HSPICE parameters.   138

6.10 The novel pull-up circuit combined with; (a) a novel pull-down circuit, and (b) a pull-down circuit similar to the one in [6].   139

6.11 The average delay of the circuits in Fig.6.12 and Fig.6.3(a) compared to CMOS for the $(0.2\mu m, 2V)$ technology   140

6.12 Power vs C_L for the circuits of Figures 6.12 and 6.3(a) compared to CMOS for the $(0.2\mu m, 2V)$ technology.   141

6.13 Delay vs supply voltage of circuits 6.12(a) and the CMOS buffer for the three BiCMOS technologies.   142

6.14 An AND gate implemented using the novel circuit technique.   143

6.15 The speed-up and area ratio between the novel BiCMOS AND and the CMOS NAND for the $(0.5\mu m, 3.3V)$ technology and as a function of the Fan_{in}.   145

6.16 A Master-Slave latch implemented using the novel circuit technique.   146

6.17 The write and total delays of the BiCMOS and CMOS latches vs the supply voltage.   147

Chapter 7

7.1 Some of the reported dynamic low-power ECL circuits; (a)The AC-PP-ECL of [6], (b) The AC-CS-APD-ECL of [7], (c) The FPD-ECL of [9], and (d) The active push-pull ECL of [8]   156

7.2 The two Bipolar current-mode drivers ; a) The ECL driver, and b) the CML driver.   157

7.3 The level shifting circuits; a) The conventional Emitter-Follower circuit, and b) The novel DAPD circuit. 158

7.4 The two current source implementations; a) The conventional current source, and b) The novel DCCP current source. 159

7.5 The collector current of Q_2 in fig.2(b) during the output pull-down and pull-up. 160

7.6 The tail current produced by the novel DCCP current source circuit during pull-up and pull-down output transitions. 163

7.7 The total power (Ptot) and the power withdrawn from VDD (Pvdd) vs frequency for the CML driver with the dynamic current source (Dyn-CML) and the conventional CML driver. 164

7.8 The micrograph of the CML driver with the new DCCP current source. 165

7.9 The measured output waveform of a CML driver with the new DCCP current source at 1.5 GHz. The output is 50Ω terminated to 3.3V (6db attenuation at the input of the sampling scope). 166

7.10 The universal input buffer (UIB) circuit. 168

7.11 The two reference circuits in the universal receiver. 169

7.12 Simulation output waveforms of the UIB for two termination voltages (2V and 5V) at 1GHz and an ECL gate at the output. 170

7.13 The maximum frequency of operation (Fmax) of the UIB vs the tail current (I_{ss}). 171

7.14 The layout of the Vref generator circuit. 172

7.15 Measurement results showing the output of the Vref generator vs temperature for different termination voltages. 173

7.16 Measurement results showing VT - Vref vs VT at room temperature. The two sets of data are from two different wafers fabricated in separate runs. 174

7.17 The outputs of the load control circuit (V1 and V2) vs the termination voltage VT. 175

7.18 The layout of the UIB circuit. 176

7.19 The micrograph of the universal transceiver test circuit. 177

7.20 The measured output waveforms of an ECL driver being driven by the new receiver at 1 GHz. The input signal to the receiver is terminated to 5V (attenuation at the sampling scope inputs is 20db). 178

List of Figures xvii

7.21 The measured output waveforms of the new receiver test structure at 1 GHz with 4V input signal termination (attenuation at the sampling scope inputs is 20db). 179

7.22 The measured output waveforms of the new receiver test structure at 1 GHz with 3V input signal termination (attenuation at the sampling scope inputs is 20db). 180

7.23 The measured output waveforms of the new receiver test structure at 1.5GHz with 2V input signal termination (20 db attenuation at the sampling scope inputs). 181

7.24 The first version of the universal output driver (UOD1). 182

7.25 The second version of the universal output driver (UOD2). 183

7.26 The maximum frequency of operation of the two UODs vs VT. 185

7.27 Schematic of the test structures of the two UODs. 186

7.28 The measured output waveforms of the two UODs with a 5V termination voltage and a 25 MHz input frequency. 187

7.29 The measured output waveforms of the two UODs with a 2V termination voltage and a 25 MHz input frequency. 188

7.30 The measured output waveforms of the two UODs at 500 MHz and a VT of 5V. The above results were produced from a different die than the one used for Figures 7.28 and 7.29 results. 190

7.31 The measured output waveform of the UOD1 at 1 GHz and 5V termination (6db attenuation at the sampling scope input). 191

LIST OF TABLES

Chapter 1

Chapter 2

2.1 Key Device Parameters for $0.8\mu m$ CMOS (in the BiCMOS Process). — 16

Chapter 3

3.1 Partial product selection. — 36
3.2 Partial product generation process. — 36
3.3 Partial product generation relations. — 37
3.4 Equivalent capacitances in simplified circuit — 46
3.5 Physical dimensions of full adders — 57
3.6 Measured results of optimized full adders — 58
3.7 Optimized transistor sizes for double rail mux (Figure 15(b)) — 58
3.8 Add cell performance — 70
3.9 Explanation of the four 6x6 multiplier implementations. — 71
3.10 6-bit multiplier performance — 73

Chapter 4

4.1 Delay and power consumption of the register file. — 93

Chapter 5

5.1 The key device parameters of the ($0.35~\mu m$, 3.3V) BiCMOS technology. — 100
5.2 The truth table of the four outputs of the W-OR pre-decoder. — 107

Chapter 6

6.1 HSPICE parameters of the three generic BiCMOS technologies. 134

Chapter 7

7.1 NT's (0.8 μm, 5V) BiCMOS Technology parameters. 155

7.2 The effects of input signals level shifting on the different performance parameters of the ECL and CML drivers. 161

PREFACE

Conventionally, low-power circuit design was only associated with very few niche applications such as wrist watches, pocket calculators, pacemakers, and some integrated sensors. However, the low-power design paradigm is becoming the norm for all high performance applications, for the power is emerging as the most important single design constraint. While designers might have different reasons for lowering the the power consumption depending on their targeted application, minimizing the overall system power has become everyone's number one daunting task. The reasons for that can be summarized in two words ; *portability* and *reliability*.

Many portable systems are being developed everyday; systems such as laptop and notebook computers and hand-held communication devices that withdraw their power from a battery. The low-power requirements for these portable systems stem from the need to both reduce the weight of the battery and extend its life before it needs re-charging. This is to enable the users to easily carry these systems with them for long periods of time. Also, users are always demanding higher performance, smaller sizes, lighter weights, and longer periods of operation out of these systems which all lead to the necessity of power reduction.

Reliability issues became a major concern for high-speed systems designs where high amount of heat is generated due to the higher clock and/or I/O frequencies. This would either imply a much shorter lifetime for these systems, less integration (which translate to lower performance and higher cost), or expensive fancy packaging and cooling techniques. This problem is more evident in high-performance systems such as the microprocessors used in desktop computers and engineering workstations, high-density static memories (SRAMs), and telecommunication switches. Core clock frequencies of state-of-the-art microprocessors are exceeding 200 MHz and ever increasing, while the number and frequency of I/Os are also increasing. For the state-of-the-art, I/O dominated telecommunication switches, where I/O frequencies are very high more I/Os mean lower overall system costs. Switches with 300+ I/Os on them are already in production and the need for higher number of I/Os is always there and persistent. The major limitation on the number of I/Os is the reliability problem that arises from the extreme heat generated by these I/Os. Again, lower-power I/O designs will mean more I/Os and/or less packaging and cooling cost, hence lower overall system cost. Reliability issues are have also became the major limitation for increasing the density of high-performance SRAMs such as ECL BiCMOS SRAMs.

In this book several novel high performance digital circuit designs that emphasis low-power and low-voltage operation are presented. These circuits represent a wide range of circuits that are used in state-of-the-art VLSI systems and hence serve as good examples for low-power design. Each chapter contains a brief introduction that serves as a quick background and gives the motivation behind the design. Each chapter also ends with a summary that briefly explains the contributions contained therein. This makes the book very readable. The reader can skim through the chapters very quickly to get a feel for the design problems presented in the book and the solutions proposed by the authors. Examples of circuits used in systems where low-power is important from reliability and portability point of views (such as general purpose and DSP processors) are presented in Chapters 2,3, and 4. Chapters 5 and 7 give examples of circuits used in systems where reliability and more system integration are the main driving forces behind lowering the power consumption. Chapter 6 give an example of a general purpose high-performance low-power circuit design.

M. S. Elrabaa
I. S. Abu-Khater
M. I. Elmasry
Waterloo, Ontario
Canada

ACKNOWLEDGEMENT

All praise is due to God, to whom we owe all gratitude.

We would like to thank our families whose support and encouragement was very determinantal to us throughout the writing of this book.

We owe a debt of gratitude to our colleagues and friends at the university of Waterloo for providing us with their valuable support, feedback and useful discussions.

A special thanks to NorTel for supporting in part some of the work presented in this book (Chapter 7).

The research work of the authors reported in this book has been supported in part by MICRONET, ITRC, CMC, BNR and NorTel. This support is greatly appreciated.

ADVANCED LOW-POWER

DIGITAL CIRCUIT TECHNIQUES

1
LOW-POWER VLSI DESIGN

In this introductory chapter, a brief description of the power estimation and evaluation for digital circuits is given in the next section. This would serve as a back ground for the subsequent chapters. Also, the impact of the new power-concious design philosophy on the process design and semiconductor technology is outlined. Finally, an overview of the book's main chapters is provided.

1.1 POWER ESTIMATION AND EVALUATION FOR DIGITAL VLSI CIRCUITS

Power dissipation in digital circuits can be categorized into two categories [1]; static power P_{DC} and dynamic power P_{DYN}. Furthermore, there are two sources for static power; power due to short circuit currents (currents flowing directly from V_{DD} to GND), P_{SC} and/or power due to leakage currents of reversed-biased PN junctions (diodes) and sub-threshold MOS current transport, P_S. Hence the general power equations for digital circuits can be summarized as follows :

$$P_{total} = P_{DYN} + P_{DC} + P_S \quad (1.1)$$

$$P_{DYN} = \alpha.(C_L.V_{DD}^2.f) \quad (1.2)$$

$$P_{SC} = I_{SC}.V_{DD} \quad (1.3)$$

$$P_S = I_S.V_{DD} \quad (1.4)$$

Where C_L is the load capacitance, α is the probability that a switching transition occurs, f is the clock frequency, V_{DD} is the supply voltage, I_{SC} is the short circuit current and I_S is the leakage (static) current. Short circuit power is the major source of power for current-mode circuits such as ECL and CML where I_{SC} flows all the time unlike pull-up/pull-down static or dynamic CMOS circuits where it flows only during input switching. The dynamic power is the major source of power for pull-up/pull-down circuits such as static or dynamic CMOS. However, for low-power applications, all power components have to be considered and minimized.

The dynamic power for a complex gate cannot be estimated by the simple expression $C_L V_{DD}^2 f$, because it might not always switch when the clock is switching. The switching activity determines how often this switching occurs on a capacitive node. For N periods of $0 \to V_{DD}$ and $V_{DD} \to 0$ transitions, the switching activity α determines how many $0 \to V_{DD}$ transitions occur at the output. Assuming that the gate does not experience glitching, the activity α represents the probability that a transition $0 \to V_{DD}$ will occur during the period $T = 1/f$. Where f is the periodicity of the inputs of the gate. So the switching activity is a complex function of several factors including: type of logic function, logic style, circuit topology and the sequencing of operation. For example an XNOR gate has a greater transition probability than a NAND gate since the XNOR has 50% 1's and 50% 0's in its truth table while the NAND gate has only one 1. This is true for both circuits regardless of the number of inputs. Also, dynamic circuit implementations in general have higher switching activities than static implementation due to the need to precharge [1].

1.2 LOW-POWER IMPACT ON PROCESS AND TECHNOLOGY

The low-power system concerns has affected both the choice and the design of process technology that implement these systems. For the I/O intensive telecommunication applications, the technology the technology of choice is BiCMOS. This allows the designer to concentrate the power where it is needed the most in high-speed blocks, while using the low-power moderate speed CMOS in other blocks. For state-of-the-art microprocessors with less I/O speed requirements and high logic density, where the supply voltage need to be as small as possible to keep the power consumption to a minimum, CMOS is the technology of choice.

Low-Power VLSI Design

The quest for low power has lead process and device designers to reduce the voltage very aggressively as they scaled down MOS devices. This meant ultra-thin MOS gates had to be implemented. Also, in order to increase the speed of the scaled CMOS devices under these low voltages, the threshold voltage had to be decreased. This in turn pushed for new processes that reduces sub-threshold currents and junction capacitances such as the shallow junction well technique [3]. New process design methodologies for deep-submicron CMOS technologies with sub-1V voltage supplies were developed [2]. Interest in CMOS SOI (Silicon-on-Insulator) processes has re-surfaced due to their reduced junction capacitances (and hence power).

As for Bipolar devices self-aligned technologies that tremendously reduced the size and parasitics and pushed the F_T's of silicon BJTs to 60+ GHz. The reduced capacitances ment lower currents are required to achieve higher frequencies.

1.3 THIS BOOK

This book is dedicated to advances in the low-power design arena. Covering several practical design examples, it emphasises circuit robustness, efficiency, and ease-of-design. It also introduces the reader to practical design issues and concerns such as circuit reliability, reproducability, and overall system cost. Targeting university students (both senior undergraduates and postgraduates) and industry VLSI sub-systems designers, it shows how the high-power problem of VLSI sub-systems can be tackled on both the circuit and architectural levels.

In Chapter 2 a low power VLSI Adder implementation is presented. Adders are a major building block for microprocessors and DSP processors. The selected architecture is explained and two circuit implementations are provided. A novel method for optimizing the architecture is demonstrated as well as the simulation and layout strategies. Experimental results from fabricated circuits are also presented.

Chapter 3 give another example of a major microprocessor and DSP block, the parallel multiplier. The different parallel multiplier architectures are reviewed with emphasis on their low-voltage low-power capabilities. The selected architecture and simulation strategy is explained. The design of the different sub-blocks in the selected architecture is demonstrated. A 6-bit multiplier design with different logic styles is presented and the performances of the different implementations are compared. The layout strategy of that design is also shown.

Another important block in processors is the register file. Chapter 4 shows the design of a modern low-voltage low-power register file. The architecture is explained, and the design of the different sub-blocks targeting high-speed low-power performance is presented. A complete 32X32 register file design and simulation is demonstrated.

In Chapter 5, novel low-power BiCMOS SRAM circuit techniques are introduced with applicability towards 16 Mb SRAMs with targeted access times of less than 5 nS. The speed and power optimization of a 16 Mb SRAM front-end is presented. Novel front-end circuits, word-driver and decoder, and a latched sense-amplifier are described and compared to conventional circuits with similar functionality. Their superior performance in terms of speed and power is demonstrated.

A novel full-swing low-voltage low-power BiCMOS circuit technique for on-chip drivers is presented in Chapter 6. The concept of operation is explained and verified using transient device simulations. Different logic gates with various complexity that are used as digital building blocks were implemented using the new circuit technique. Their performance is compared to that of CMOS gates with similar functionality. The comparison includes delay, power, and delay versus supply voltage for equal input capacitance. A brief note on the design procedure of the new circuit is provided.

In Chapter 7 low-voltage-swing inter-chip transceivers are presented. Several CML/ECL dynamic circuit techniques that increases the speed and/or reduces the power of conventional CML/ECL circuits are described. Their performance is substantiated using results from simulations and experimental measurements of circuits fabricated in a $0.8 \mu m$ BiCMOS technology. A new high-speed low-power universal transceiver (Receiver/Driver) that can operate with a large range of termination voltages without the need for external reference voltage is also presented. The design and performance of each block in the transceiver is discussed in details. Results from measurements performed on fabricated circuits that verify the performance are also provided.

REFERENCES

[1] A. Bellaouar and M. I. Elmasry, "Low-Power Digital VLSI Design Circuits and Systems", Kluwer Academics Publications, MA, 1995.

[2] H. Oyamatsu, K. Kinugawa, and M. Kakumu, "Design Methodology of Deep Submicron CMOS Devices for 1 V Operation," Symposium on VLSI Technology Tech. Dig., pp. 89-90, 1993.

[3] H. Yoshimura, F. Matsuoka, and M. Kakumu, "New CMOS Shallow Junction Well FET Structure (CMOS-SJET) for Low Power-Supply Voltage," International Electron Devices Meeting Tech. Dig., pp. 909-912, December 1992.

2

LOW-POWER
HIGH-PERFORMANCE ADDERS

2.1 INTRODUCTION

The scaling of the CMOS channel length to below 0.5 μm and increasing of the chip density to the ULSI range have placed power dissipation on an equal footing with the performance as a figure of merit in digital circuit design. Portability and reliability [1] have also played a major role in the emergence of low-power, low-voltage, digital circuit designs. The need to extend the battery life, to have inexpensive packaging and cooling systems, and to reduce the weight and size of the equipment were the driving forces in this regard.

Reducing the power dissipation of arithmetic operations while keeping the performance unaffected, is indispensable for digital signal processing (DSP) , reduced instruction set computers (RISCs), microprocessors, etc. As an example, high-speed low-power adders are essential as arithmetic blocks.

This Chapter explores the circuit and architecture techniques for a 32-bit adder macro targeting low-voltage/low-power applications. The proposed adder is based on the Conditional Sum Addition (CSA) algorithm [2] combined with Carry Select (CS) . The main contribution is in realizing the CSA adder using a low-power Complementary Pass-Transistor Logic (CPL-like) circuit style. It will be shown that this implementation is better than its TG style counterpart [3].

The performance of this adder is compared to different 32-bit adders implemented in 0.8 μm CMOS(in BiCMOS) technology in the range of 3.3 to 1.5 V, using different architectures (i.e. Carry Look Ahead (CLA), CS, manchester) with different circuit styles (i.e. conventional static CMOS, Transmission Gate

(TG), CPL [4], and Double Pass-transistor Logic (DPL) [5]). Unlike other comparisons [6, 7], this work explores the minimum size transistor design and then optimizes the delay of the critical path, and hence keeps the power dissipation low. Thus, two sets of comparisons among all 32-b adder architectures are presented. The first is for a minimum transistor size design and the second is for optimized circuit performance.

In Section 2.2 the CSA architecture is described along with the critical path delay analysis. The circuit implementation of the CSA adder is discussed in Section 2.3. Section 2.4 presents the effect of block sizes and staging on the performance of the CSA adder. Sections 2.5 and 2.6 discuss the simulation strategy and the performance comparison of the CSA adder to other adder architectures, respectively. The layout strategy is presented in Section 2.7 and that is followed by concluding remarks in Section 2.8.

2.2 ARCHITECTURE

A 4-bit block diagram of the Conditional Sum Adder (CSA) is shown in Figure 2.1(a). The adder is composed of conditional sum cells, and a block of 2 to 1 multiplexers (MUXs). Figure 2.1(b) shows the gate level implementation of the conditional sum cell, where S^0 and C^0 are the sum and carry corresponding to a carry in of "0", and S^1 and C^1 correspond to a carry in of "1". Figure 2.1(c) shows the implementation of the MUXs using the TG logic style.

To design an n-bit adder, one possible technique for fast operation is to use staged blocks with variable widths. In this case, all the conditional sum blocks compute their respective double sums and double output carries in parallel. The true carry-out and sum of each block are then selected by the carry-in generated by the previous stage. The architecture at the block level uses a CS-like technique.

Thus the adder circuit has two critical delay paths, τ_{Cin} and τ_{B1}. A 4-bit adder implementation example, using two blocks of size 2-bits each, is shown in Figure 2.2 for the purpose of illustration. τ_{Cin} is the delay from the input carry to the carry out [$C_{o2} \rightarrow C_{o4}$], this delay totals the sum of the output MUXs delay in each block. τ_{B1}, the longer one, is the delay from the LSB-input (B1) to the carry out [B1$\rightarrow C_{o2}$] and is composed of the carry generation delay in the conditional cell, the carry delay through the MUXs within a block and the delay of selecting the output carry to next block τ_{Cin}. Both delay

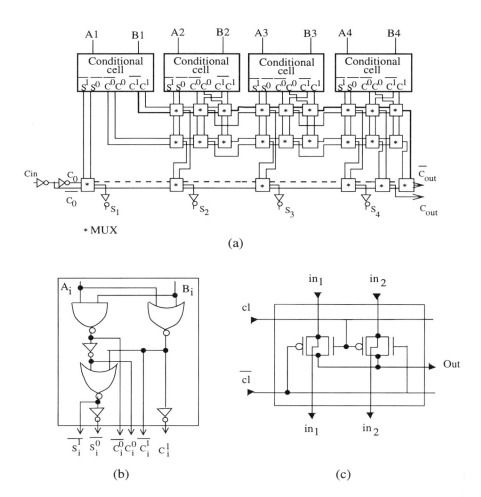

Figure 2.1 Conditional Sum with Carry Select Adder: (a) 4-bit block architecture, (b) implementation of conditional circuit (c) multiplexer TG implementation.

paths are highlighted in Figure 2.2. These propagation delays are estimated for an n-bit adder using variable staging to be:

$$\tau_{Cin} = \tau'_{MUX}\left(\frac{N}{B}\right) \tag{2.1}$$

$$\tau_{B1} = \tau_{cond} + (S_1 - 1)\tau_{MUX} + \tau_{Cin} \tag{2.2}$$

where:
N = number of bits
B = number of blocks
S_1 = size of first block
τ_{cond} = conditional cell delay
τ'_{MUX} = inter-block delay of output MUXs
τ_{MUX} = intra-block delay of MUXs within the block

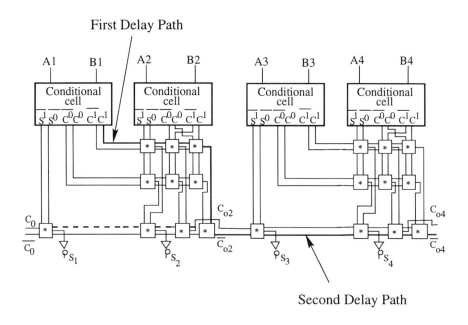

Figure 2.2 Critical delay path of 4-bit CSA adder.

The above equations show that the MUXs design and the number of blocks and sizes used in building higher order n-bit adders are crucial elements in affecting the performance and power dissipation of the circuit. Hence, optimizing the circuit should take into consideration such effects.

Low-Power High-Performance Adders

The CSA adder could be constructed using a small number of basic cells as shown in Figure 2.1. This simplifies the layout of the circuit and also the simulation and optimization. The ALU functions are also available at no extra cost as highlighted in Figure 2.1(b). This will reduce the power consumed by the extra circuitry needed to realize the ALU functions in conventional designs.

2.3 CIRCUIT DESIGN AND IMPLEMENTATION

The adder is implemented using static design, and hence this eliminates the precharging and reduces extra power dissipation by the clocking. Two circuit styles for the CSA adder are presented, the first is based on the TG style and the proposed second implementation uses a CPL-like logic style.

2.3.1 TG Implementation

The static-TG implementation was constructed using simple static logic gates for the conditional cell as was shown in Figure 2.1(b). The MUXs were implemented using the TG style as shown in Figure 2.1(c). A full discussion on the static TG implementation is presented in [3] and will not be discussed in this Chapter. It will only be used for comparisons with the new implementation. Although the static TG implementation exhibits high performance and low power compared to other adder families, it suffers from the presence of large PMOS transistors which occupy more area and contribute to the dynamic power dissipation.

2.3.2 CPL-Like Implementation

The alternative CPL-like implementation was sought for three main reasons:

1. It eliminates the use of PMOS transistors in the MUXs.
2. Using NMOS will reduce the voltage swing on the internal nodes by V_T subject to the body effect, which contributes to lowering the dynamic power dissipation.

3. Complementary signals are available which is compatible with the proposed adder architecture and eliminate the use of inverters in the conditional circuit.

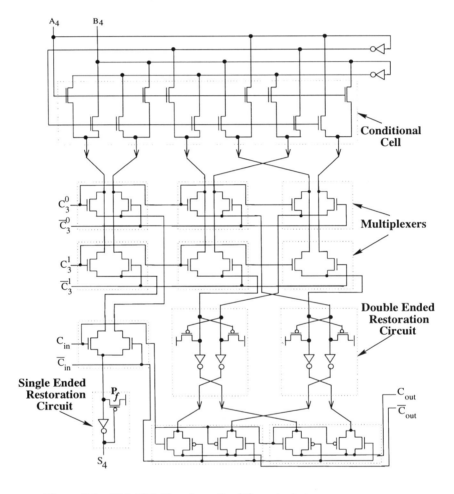

Figure 2.3 CSA-CPL-like schematic of the output stage.

The transistor implementation of the output stage (i.e. 4th-bit of Figure 2.1(a)) of the building block using the CPL-like logic is presented in Figure 2.3. The output stage is shown because it contains all elements of the design. The conditional cell and the MUXs are designed using CPL logic. The conditional circuit generates the double sum and carry and their complement with a reduced swing, and then the generated signals drive the MUXs without the need for signal restoration. The voltage drop on the internal switching capacitance nodes will be translated into savings in the dynamic power dissipation.

The selected sum is restored to full-swing using a single ended restoration circuit as shown in Figure 2.3. The feedback PMOS transistor P_f is of minimum size and is carefully designed as a function of the four NMOS transistor stages in the sum path. The feedback PMOS is needed to eliminate the static power consumption in the circuit due to the reduced voltage swing.

The MUXs are implemented in NMOS pass logic, and all signals driving the MUXs' gates are restored to full-swing by the employment of a double ended restoration circuit as shown in Figure 2.3. This is required to insure that the transistors are fully turned on.

The output MUXs of the block which selects the carry-out to drive the next block are designed using TG logic. This is required in order to of keep the output carry at full-swing. The restoration circuitry in the output stage acts as a buffer driving the output TG-MUX. It was shown in [3] that placing the buffers within the block and ahead of the output MUXs will eliminate them from being in the intra-block critical path and thus will not contribute to the overall circuit delay of the staged blocks.

The advantage of CPL-like implementation over the TG (in carry path) counterpart is the minimization of PMOS transistors in the circuit and hence reduce both the internal nodes' switching capacitance and the layout area used.

2.4 OPTIMIZATION USING VARIABLE BLOCK SIZE COMBINATION

Optimizing the adder performance is directly dependent on the block size and staging used to build the adder. The effect of the block size and staging is illustrated in the 16-bit adder example shown in Figure 2.4. The two critical path delays of the building blocks are indicated on the block, where the top and bottom numbers correspond to the two delay path of equations (2) and (1), respectively. A 16-bit adder can be built using many different staging combinations of varying block sizes, in Figure 2.4 we are presenting two solutions.

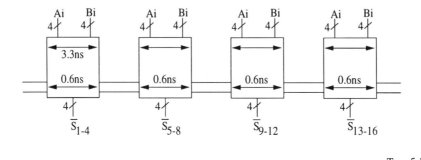

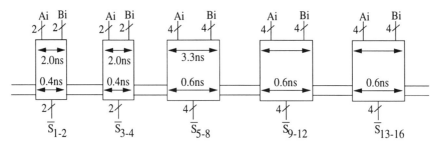

Figure 2.4 Block size effect on performance of 16-bit Adder: (a) equal block size (b) variable block sizes.

Low-Power High-Performance Adders 15

Figure 2.4(a) shows the first implementation using blocks of equal sizes, in this case 4 blocks of 4-bits each. The propagation delay is computed based on the fact that all blocks are executed in parallel, so after 3.3ns delay of the first block, the only delay is through propagating the carry to the next block, hence the delay can be computed as follows:
$$3.3ns + 3*.6ns = 5.1ns$$
It is worth while to note that the fourth block will be idle for at least 1.2ns waiting for the carry to arrive.

The idle time can be reduced if the block size and staging is changed to the second implementation shown in Figure 2.4(b). Here the blocks have variable sizes based on the staging condition that the block size is increased only when the next block is settled and ready for the carry-in when it arrives. This optimization will lead to a better performance and in this case, the optimized architecture is composed of two blocks of 2-bits each followed by three 4-bit blocks and its delay is 4.5ns.

For the case of the 32-bit adder, simulations for optimized speed have resulted in staging of block sizes of 2-2-4-4-4-8-8. This combination is used for all simulation results of the 32-bit CSA architecture throughout this Chapter.

2.5 SIMULATION STRATEGY

2.5.1 Transistor Sizing

Since a minimum size design leads to minimum power dissipation, the simulation was carried out on a minimum size transistor design for each adder architecture. The size of the PMOS used is double that of the NMOS. The design then is optimized to improve the performance of the circuit. This strategy will lead to power savings, since only the critical delay paths are optimized.

2.5.2 Power Estimation

In order to give an accurate estimation of the power dissipation, we have constructed the 32-bit adders using modules of 4-bit blocks except for the CS and CSA implementations. The CS uses a 4-4-7-9-8 staging combination of block sizes. The CSA uses the combination mentioned in the previous Section. The power dissipation of a certain adder architecture is estimated by summing the average power of each block in the circuit. The average power of a block is computed by averaging it over all possible input patterns. An estimated load at the output of the module is included.

2.6 CIRCUIT PERFORMANCE

Table 2.1 Key Device Parameters for 0.8μm CMOS (in the BiCMOS Process).

	NMOS	PMOS
L_G	0.8μm	0.8μm
I_{ds}	1.8 mA	0.8 mA
@ $V_{DS} = V_{GS}$		= 3.3V, W = 10 μm
T_{ox}	175 Å	175 Å
V_T	0.80V	-0.90V

The adder circuits are simulated using the 0.8 μm CMOS (in BiCMOS) technology, the simulator used is HSPICE [8]. Table 2.1 lists the key technology parameters for the process used.

The CSA adder architecture is compared to the CLA, CS and manchester adder architectures using conventional static CMOS, TG, CPL and DPL circuit styles. Figure 2.5 shows the critical carry path circuits of the various architectures used. The various adders are named to reflect the architecture and circuit style as shown in Figure 2.5. For all the CLA adder architectures, P and G are the global generate and propagate signals given by:

$$P = P_{i+3}P_{i+2}P_{i+1}P_i \quad (2.3)$$

$$G = G_{i+3} + P_{i+3}G_{i+2} + P_{i+3}P_{i+2}G_{i+1} + P_{i+3}P_{i+2}P_{i+1}G_i \quad (2.4)$$

where: $P_i = A_i \oplus B_i$, $G_i = A_i \cdot B_i$, A_i and B_i = the ith bits of the numbers to be added.

Figure 2.6 shows the energy versus delay for the minimum-size 32-bit adders at 3.3 V. It is obvious that both of the CSA (i.e, CPL-like and TG) implementations consume less energy and provide a better speed performance than any other architecture. The CSA-CPL implementation consumes 22% less energy at approximately the same performance compared to the CSA-TG one. In addition, the CSA-CPL circuit does not exhibit any DC current consumption due

Low-Power High-Performance Adders 17

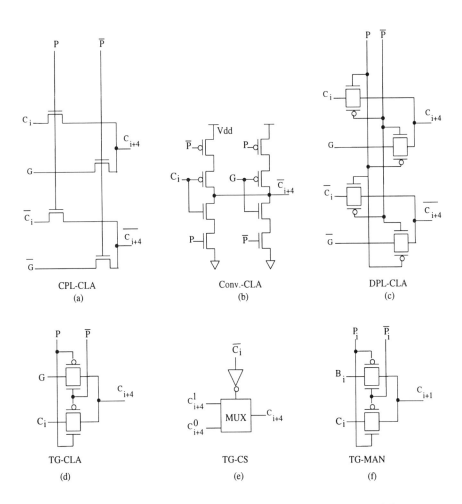

Figure 2.5 Critical carry delay path of: (a) conventional-CLA, (b) CPL-CLA, (c) DPL-CLA (d) TG-CLA, (e) TG-Carry Select (CS), (f) and TG-manchester

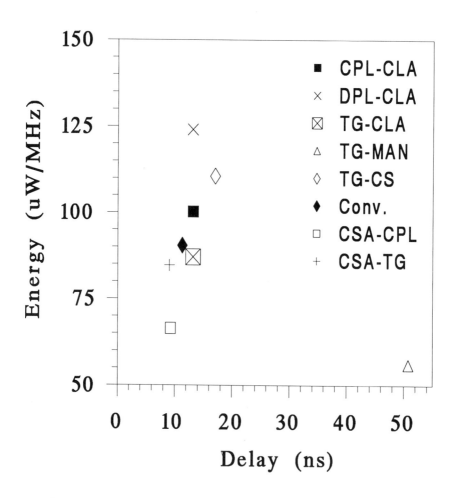

Figure 2.6 Energy versus delay for the minimum transistor size adder (V_{DD}=3.3 V).

to the use of the restoration circuits and feedback PMOS transistors. Compared to the conventional static CLA implementation, the CSA-CPL consumes 27% less energy and enhances the speed by 18%. Compared to the TG-manchester, the CSA-CPL consumes 15% more energy but provides a delay reduction factor of 5.5.

Figure 2.7 shows the energy versus delay for the optimized 32-bit adders, again the CSA-CPL implementation consumes 22% less power than the conventional-CLA and 34% faster. When compared to the TG-manchester adder the CSA-CPL implementation consumes almost the same amount of energy but it outperforms it in terms of speed by a factors of 5. The optimized CSA-CPL implementation consumes 45% less energy than the optimized CSA-TG circuit with a speed reduction of 17%.

In order to study the sensitivity of the circuit performance to the scaling of the supply voltage, we have simulated the optimized circuits with different supply voltages. The threshold voltage was scaled with the supply using the relation, $V_T = 0.2V_{DD}$. Figure 2.8 shows the delay versus supply voltage scaling for the optimized adders which demonstrates that both of the CSA implementations are faster than any other implementation down to 1.5V.

Figure 2.9 shows the power dissipation versus the supply voltage scaling at 20 MHz, the CSA-CPL implementation provides the lowest power dissipation, comparable to none other than the TG-manchester implementation which has a low-power consumption and an extremely high propagation delay.

The performance of the CSA-CPL circuit is slightly affected compared to that of the CSA-TG, this is due to the adding of the restoration circuitry in the critical path. But the overall power-delay product is reduced by 35% using the CPL-like implementation.

The effect of scaling the supply voltage from 3.3 to 1.5 V on the CPL-like implementation has produced a power saving factor of 4.4 and a speed degradation factor of 1.8. This will translate into a power-delay product saving of a factor of 2.4.

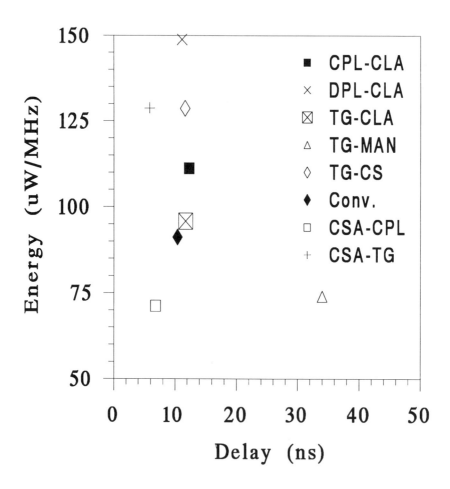

Figure 2.7 Energy versus delay for the optimized transistor size adder (V_{DD}=3.3 V).

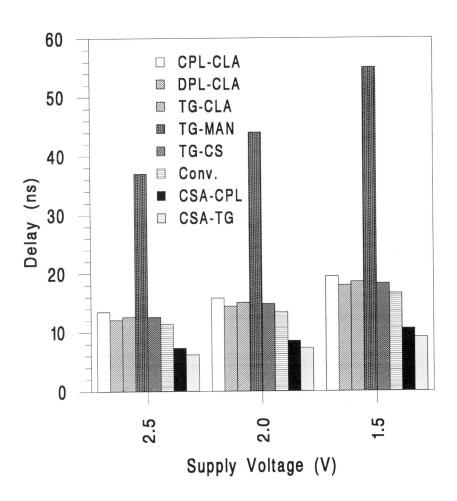

Figure 2.8 Effect of V_{DD} scaling on the adders' delay.

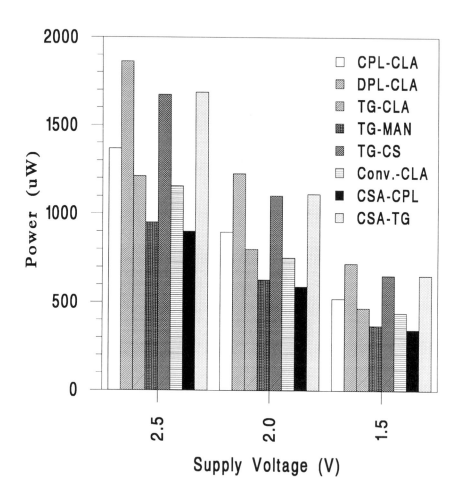

Figure 2.9 Effect of V_{DD} scaling on the adders' power dissipation at 20 MHz.

2.7 LAYOUT STRATEGY

The CSA-CPL circuit was laid out for fabrication in order to perform an experimental evaluation of the circuit. The technology used was the 0.8μm CMOS (in BiCMOS) process. In constructing the circuit layout, consideration was given to minimizing the critical path routing capacitance.

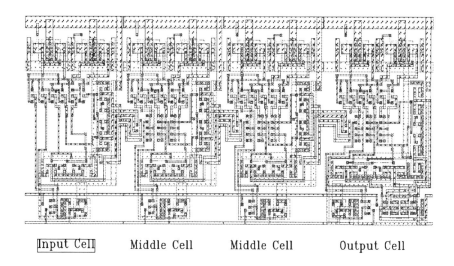

Figure 2.10 Layout illustration of a 4-bit CSA-CPL adder in 0.8 μm technology.

The layout explored the modularity feature in the CSA architecture by designing three basic 1-bit adder cells; the input, middle, and output. Then using a standard cell approach, the n-bit block can be realized by a combination of one input, (n-2) middle, and one output cells. Then the blocks are assembled to form the desired adder size. A 4-bit illustration example of the layout strategy is given in Figure 2.10.

2.8 EXPERIMENTAL RESULTS

The photomicrograph of the test chip and that of the 32-bit CSA-CPL implementation are shown in Figure 2.11 (a) and (b), respectively. The delay is measured from B_1 to the output carry C_{32}. A separate supply voltage was used

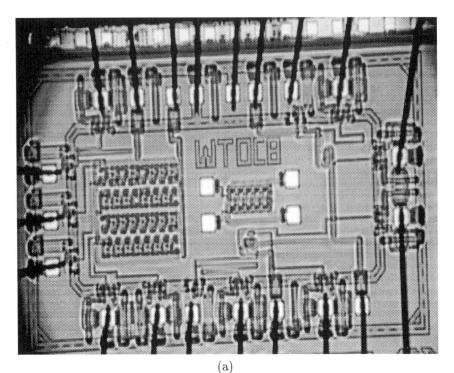

(a)

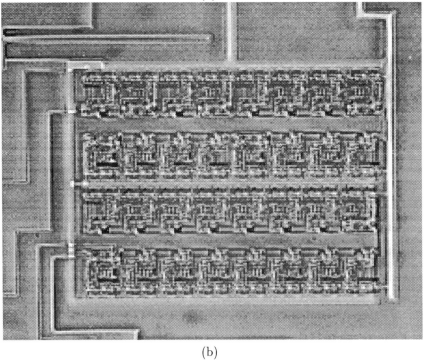

(b)

Figure 2.11 Photomicrograph of (a) Test Chip and (b) 32-bit CSA-CPL adder.

to measure the RMS current of the adder. The testing is carried out with a 1MHz pulse at 3.3 V supply voltage.

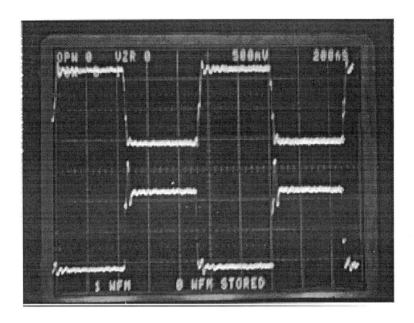

Figure 2.12 CSA-CLA functionality test; input waveform B_1 (bottom) and output S_{32} (top).

The adder functionality is shown in Figure 2.12 where the bottom waveform is the B_1 input and the top waveform is the S_{32} output. This correspond to an input vector where the A_i's are all set to logic "0", B_i's are all set to logic "1" except for B_1 which is switching, and C_{in} is at logic "1". This vector allows us to compute the critical delay path, since the evaluation of the sum depends on the carry from the previous block.

The rise time and fall time delay measurements are shown in Figure 2.13 (a) and (b) respectively. An average delay of 9.5ns is obtained which is 30% larger than the simulated results. The 30% increase in the delay is attributed to the routing capacitance which was not included in the simulation. This result is consistent with the simulation from the layout extraction with parasitic capacitances.

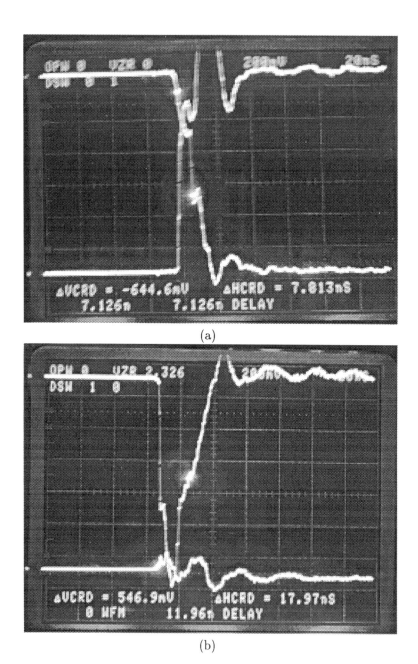

Figure 2.13 50% time delay measurement for the CSA-CPL adder between B1 and inverted S32 (a) rise time (b) fall time.

The measured RMS current is 17.75 μA at a frequency of 1MHz. This corresponds to a power dissipation of 58.6 μW. This result is in agreement with the simulation from the layout extraction with parasitic capacitances.

2.9 SUMMARY

A novel 32-bit adder has been designed using a CSA architecture combined with a carry select and CPL-like circuit implementation. The new implementation outperforms all other implementations in terms of power and speed, a power savings of more than 22% over conventional-CLA and a speed increase of 34% was achieved at 3.3V. These power and speed advantages are maintained down to a 1.5 V power supply. Compared to the CPL-CLA and DPL-CLA implementations, the CSA-CPL adder outperforms both of them by a power saving of 36% and 52% and a speed enhancement of 44% and 38%, respectively. The circuit is also applicable for low-voltage applications down to 1.5V.

Test results confirmed the above results in terms of speed and power dissipation. It also proved that the CPL circuits are a viable solution for low-power, low-voltage operation.

REFERENCES

[1] Anantha P. Chandrakasan, et al, "Low-Power CMOS Digital Design", IEEE J.S.S.C, Vol. SC-27, No. 4, pp. 473-484, April 1992.

[2] J. Sklansky, "Conditional-Sum Addition Logic", IRE Trans. on Electro nic Computers, June 1960, pp.226-231.

[3] I.S. Abu-Khater, R.H. Yan, A. Bellaouar and M.I. Elmasry, "A 1-V Low-Power High-Performance 32-Bit Conditional Sum Adder", 1994 IEEE Symposium onLow Power Electronics, Oct 10-12, 1994, San Diego, pp. 66-67.

[4] K. Yano, et al., "A 3.8-ns CMOS 16x16 Multiplier Using Complementary Pass-Transistor Logic", IEEE J. Solid-State Circuits, vol. SC-25, no. 2, pp. 388-394, April 1990.

[5] M. Suzuki, et al., "A 1.5-ns 32-b CMOS ALU in Double Pass-Transistor Logic", IEEE J. Solid-State Circuits, vol. SC-28, no. 11, pp. 1145-1151, November 1993.

[6] T.K. Callaway and E.E. Swartzkander, "Estimating the Power Consumption of CMOS Adders", 11th IEEE Symposium on Computer Arithmetics, 1993, pp. 210-216.

[7] C. Nagendra et al., "Power-Delay Characteristics of CMOS Adders", IEEE Trans. on Very Large Scale Integration Systems, Vol. 2, No. 3, Sept. 1994, pp. 377-381.

[8] HSPICE Version H92, Meta-Software, Inc, 1992.

3
LOW-POWER HIGH-PERFORMANCE MULTIPLIERS

3.1 INTRODUCTION

Digital signal processing (DSP) is the technology at the heart of the next generation of personal communication systems. Most DSP systems incorporate a multiplication unit to implement algorithms such as convolution and filtering. In many DSP algorithms, the multiplier lies in the critical delay path and ultimately determines the performance of the algorithm. Figure 3.1 shows an implementation of the discrete cosine transform, an algorithm widely used in image and video compression. This particular implementation requires 32 convolutions and 8 additions [1]. Thus, improving the throughput of this algorithm requires a high-performance multiplier. Traditionally in order to achieve high performance multipliers, parallel addition of the partial products is used along with reducing the technology feature size.

In the past, most of the research and design efforts have focused on increasing the speed and throughput of DSPs. As a result, present technologies posses computing capacities that allow the realization of computationally intensive tasks such as speech recognition and real time digital video. However, the demand for high-performance portable systems incorporating multimedia capabilities has elevated the design for low-power to the forefront of design requirements in order to maintain reliability and provide longer hours of operation.

In this Chapter, we present circuit design techniques to implement high performance and low-power multipliers. One of the main contributions is in performing an extensive study of existing and novel FA circuits for low-power applications. This resulted in a low energy FA called the CPL-TG FA. Another contribution is made in implementing the Booth encoder in low-power

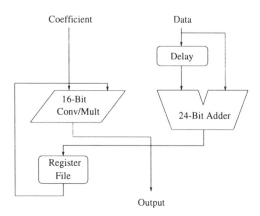

Figure 3.1 Discrete Cosine Transform

CPL circuit logic style. The modified Booth architecture is used as a test vehicle. The summing stage adder used in the multiplier simulation is the CSA adder presented in Chapter 2 and is not discussed in this Chapter. Since the adder is a stand alone application and is independent from the multiplier design, devoting a seperate Chapter for it is justified. Finally the goal of this work is to devise novel low-power, low-voltage circuit techniques which can be used in implementing multipliers in general. A Booth multiplier is used to demonstrate the performance of such circuits. Hence, comparison with other multipliers is not included.

In Section 3.2 a review of some of the existing parallel multipliers is presented. A comparison of the various architectures is also given. In Section 3.3 the circuit architecture and simulation method used is described. In Sections 3.4 - 3.7 the design and optimization of the multiplier building block are discussed. In Section 3.8 the simulation results for a 6-bit modified Booth multiplier is presented. The impact and tradeoffs of the logic circuit style used is also discussed.

3.2 REVIEW OF PARALLEL MULTIPLIERS

In this Section, several parallel multiplier algorithms which have been used in VLSI are briefly presented.

3.2.1 Braun Multiplier

Consider two unsigned numbers $X = X_{n-1}...X_1X_0$ and $Y = Y_{n-1}...Y_1Y_0$

$$X = \sum_{i=0}^{i=n-1} X_i 2^i \quad (3.1)$$

$$Y = \sum_{i=0}^{i=n-1} Y_i 2^i \quad (3.2)$$

The product $P = P_{2n-1}...P_1P_0$, which results from multiplying the multiplicand X by the multiplier Y, can be written in the following form

$$P = \sum_{i=0}^{i=n-1} \sum_{j=0}^{j=n-1} (X_i Y_j) 2^{i+j} \quad (3.3)$$

Each of the partial product terms $P_k = X_i Y_j$ is called a summand. Figure 3.2(a) shows an example of a 4×4 multiplication. The summands are generated in parallel with AND gates. Figure 3.2(b) shows the Braun's array multiplier [2]. Such a multiplier of $n \times n$ requires $n(n-1)$ adders and n^2 AND gates.

The delay of such a multiplier is dependent on the delay of the full-adder cell and the final adder in the last row. In the multiplier array, a full-adder with balanced carry and sum delays is desirable because sum and carry signals are both on the critical path. For large arrays, the speed and power of the full-adder are very important.

3.2.2 Baugh-Wooley Multiplier

It was noted that a Braun multiplier performs multiplication of unsigned numbers. The Baugh-Wooley technique [2] was developed to design regular direct multipliers for two's complement numbers. This direct approach does not need any two's complementing operations prior to multiplication. The schematic circuit diagram of a 4×4 two's complement multiplier based on Baugh-Wooley's algorithm is shown in Figure 3.3. In this scheme $n(n-1) + 3$ full-adders are required. So for the case of $n = 4$ the array needs 15 adders. When n is relatively large, the final adder stage in the multiplier array can be implemented with the CLA or CSA addersCPL-like discussed in Chapter 2.

This type of multiplier is suitable for applications where operands with less than 16-bits are to be processed. Applications for such a multiplier are, for

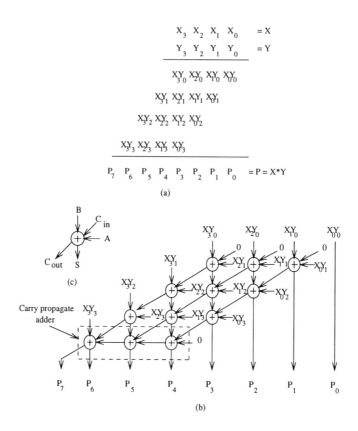

Figure 3.2 (a) Partial products of a 4 × 4 unsigned integer multiplication; (b) multiplier array; (c) full-adder schematic.

example, in digital filters where small operands are used (e.g., 6, 8 and 12). For operands equal to or greater than 16-bits, the Baugh-Wooley scheme becomes area-consuming and slow. Hence, techniques to reduce the size of the array, while maintaining the regularity are required.

3.2.3 The Modified Booth Multiplier

For operands equal to or greater than 16-bits, the modified Booth algorithm [3] has been widely used. It is based on encoding the two's complement operand (i.e., multiplier) in order to reduce the number of partial products to be added.

Low-Power High-Performance Multipliers

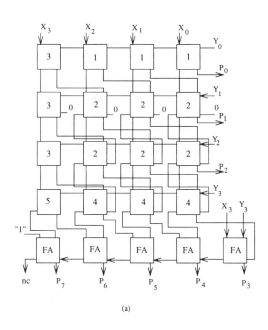

Figure 3.3 (a) 4 × 4 Baugh-Wooley two's complement regular array (FA : Full-Adder).

This makes the multiplier faster and uses less hardware (area). For example, the modified Radix-2 algorithm is based on partitioning the multiplier into overlapping groups of 3-bits, and each group is decoded to generate the correct partial product.

The multiplier, Y, in the two's complement can be written as:

$$Y = -Y_{n-1}2^{n-1} + \sum_{i=0}^{i=n-2} Y_i 2^i \tag{3.4}$$

It can be rewritten as follows

$$Y = \sum_{i=0}^{i=n/2-1} (Y_{2i-1} + Y_{2i} - 2Y_{2i+1}) \cdot 2^{2i} \quad \text{with } Y_{-1} = 0 \tag{3.5}$$

In this equation, the terms in brackets have values in the set $\{-2, -1, 0, +1, +2\}$. The encoding of Y, using the modified Booth algorithm, generates another number with the following five signed digits, -2, -1, 0, +1, +2. Each encoded digit in the multiplier performs a certain operation on the multiplicand, X, as illustrated in Table 3.1.

Table 3.1 Partial product selection.

Y_{2i+1}	Y_{2i}	Y_{2i-1}	Recoded digit	Operation on X
0	0	0	0	$0 \times X$
0	0	1	+1	$+1 \times X$
0	1	0	+1	$+1 \times X$
0	1	1	+2	$+2 \times X$
1	0	0	-2	$-2 \times X$
1	0	1	-1	$-1 \times X$
1	1	0	-1	$-1 \times X$
1	1	1	0	$0 \times X$

The bits of the multiplier (Y) are partitioned into groups of overlapping 3-bits, each group permits the generation of a certain partial product. The five possible multiples of the multiplicand are generated based on the explanation given in Table 3.2.

Table 3.2 Partial product generation process.

Recoded Digit	Operation on X
0	Add 0 to the partial product
+1	Add X to the partial product
+2	Shift left X one position and add it to the partial product
-1	Add two's complement of X to the partial product
-2	Take two's complement of X and shift left one position

The generated partial product is related to the multiplicand for each encoded digit by the relationships presented in Table 3.3. PP_i is the partial product and PP_n is the sign bit of the partial product with $P_n = P_{n-1}$ when no shifting of the partial product is performed. Note that the partial product is represented on $n+1$ bits.

Table 3.3 Partial product generation relations.

Recoded Digit	Operation on X		Added to LSB
0	$PP_i = 0$	for $i = 0, \cdots n$	0
+1	$PP_i = X_i$	for $i = 0, \cdots n$	0
+2	$PP_i = X_{i-1}$	for $i = 0, \cdots n$	0
-1	$PP_i = \bar{X}_i$	for $i = 0, \cdots n$	1
-2	$PP_i = \bar{X}_{i-1}$	for $i = 0, \cdots n$	1

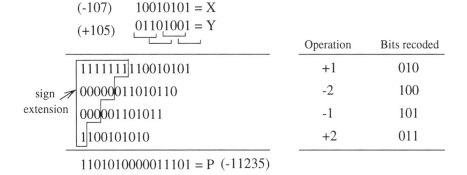

Figure 3.4 8-bit Booth multiplication example using two complement's number.

To clarify this algorithm, an example is presented in Figure 3.4. Let $X = 10010101$ and $Y = 01101001$. The encoded digits of Y are

$$01101001 \rightarrow +2 \; -1 \; -2 \; +1$$

The bits are grouped into 3-bit groups overlapping by one bit and a bit with a value of zero is added on the right side of Y as Y_{-1}. So the multiplication of two 8-bit numbers generates only 4 partial products. The number is then reduced by half. The partial product in this example is represented on 9-bits. For a correct partial product's addition, the signs are extended as shown in Figure 3.4. The shape of the multiplier is then trapezoidal due to the sign extension.

In order to make the array rectangular, and then more regular for VLSI implementation, the problem of the sign extension must be addressed. This problem is more crucial when the operand lengths are wide, where each partial product must be sign-extended to the length of the product. One such technique is shown in Figure 3.5 for the example of Figure 3.4. The basic idea is to use two extra bits in the partial product. For the first partial product, the two additional bits, PP_{n+1} and PP_{n+2} are equal to the sign bit of the partial product

$$PP_{n+2} = PP_{n+1} = PP_n \quad (3.6)$$

For the second partial product, if the first partial product was positive, then the two additional bits for this second partial product are given by the expression above, otherwise we have two cases

$$PP_{n+2} = PP_{n+1} = 1 \quad if \quad PP_n = 0 \quad (3.7)$$

and

$$PP_{n+2} = \overline{PP_{n+1}} = 1 \quad if \quad PP_n = 1 \quad (3.8)$$

So it is more interesting to use a third bit, F, as a flag to indicate whether there is, from the previous partial, a negative sign bit to be propagated. F_1 is the flag generated by the first partial product to the next one. For the example of Figure 3.5, $F_0 = 0$ (no PP before the first one), and $F_1 = F_2 = F_3 = 1$. So for the first partial product there is a sign propagation to all the others. This flag is expressed by the following Boolean equation

$$F_{j+1} = F_j + PP_{n,j} \quad (3.9)$$

where $PP_{n,j}$ is the sign bit of the jth partial product.

Figure 3.6 shows the block diagram of an $n \times n$ modified Booth multiplier. Furthermore, the figure gives an idea about the floorplan of this subsystem. It is composed of the following blocks:

- The multiplier array containing partial product's generators and 1-bit adders;

- The Booth encoder and the sign extension bits (PP_{n+2}, PP_{n+1}, F). The Booth encoder generates the five signals ($0, +1\times, +2\times, -1\times,$ and $-2\times$) for each group of 3-bit of Y; and

Low-Power High-Performance Multipliers

```
(-107)   10010101 = X
(+105)   01101001 = Y
```

	Operation	Bits recoded
¦1̄1¦110010101	+1	010
[1̄1]011010110	-2	100
[1̄1]001101011	-1	101
[0]100101010	+2	011

1101010000011101 = P (-11235)

¦⁻¦
¦_¦ Additional bits to be generated [sign extension]

☐ Additional bits generated from the previous sign and the present sign

Figure 3.5 The previous example of Figure 3.4 with simplified sign extensions.

- The final stage adder performs $2n$ bits addition.

The Booth multiplier exhibits unnecessary glitches. The main reason for glitches is due to the race condition between the multiplicand and the multiplier due to the Booth encoder.

3.2.4 Wallace Tree

By applying the Booth algorithm, the number of partial products is halved. However, for large multipliers, 32-bit and over, the number of the partial products is over 16-bit. In this case, the performance of the modified Booth algorithm is limited. One technique, to improve the performance of these multipliers, is to adopt the Wallace tree using 4-2 compressors. A 4-2 compressor accepts 4 numbers and a carry in, and sums them to produce 2 numbers and carry out. Figure 3.7(a) shows an example of such a tree on partial products of an unsigned 8×8 multiplier. Eight partial products are produced. Using 4-2 compressors, two levels of additions (stages) are needed. The final two summands are added using a fast 16-bit adder. Some zeros are added to the

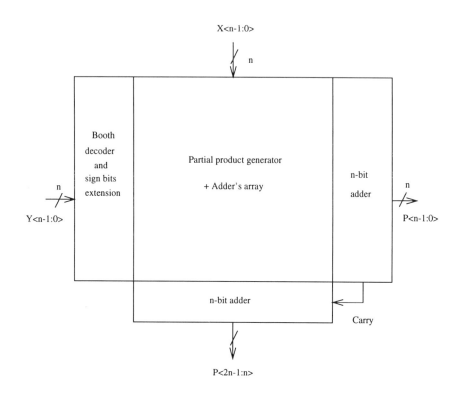

Figure 3.6 Block diagram of the $n \times n$ multiplier using a modified Booth algorithm.

array. This example shows that the bits which are not used in the 1st stage (level) jump to the next one to be combined with the ones produced by the compressors. Figure 3.7(b) shows the architecture of the 8 × 8 multiplier. For the first stage of the tree, two blocks, A and B, are required. The block A (B) of compressors group the first (last) four partial products, respectively.

To further enhance the Wallace tree multiplier, the modified Booth algorithm can be used to reduce the number of partial products by half in a carry-save adder array . One example of such combined construction is the architecture of the 32 × 32 multiplier shown in Figure 3.8. It consists of four functions: the Booth encoder , the partial product's generator , the compressor blocks, and the final 64-bit adderi . The Wallace tree is constructed with 3 stages (levels). The first stage has 4 blocks (A to D), with each block summing up 4 partial

Low-Power High-Performance Multipliers 41

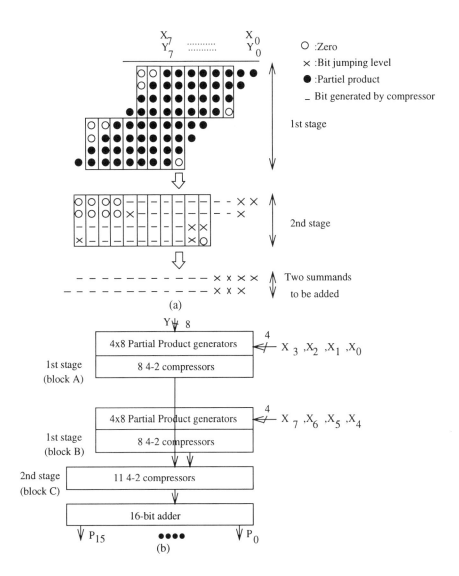

Figure 3.7 Construction of Wallace's tree for an 8 × 8 multiplier: (a) reduction of the 8 partial products with 4-2 compressors; (b) the architecture.

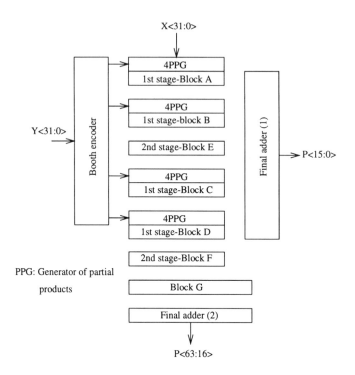

Figure 3.8 The architecture of 32 × 32 modified Booth multiplier with Wallace tree.

products among 16. The second stage sums up the 8 new generated partial products from the first stage. Hence, two blocks are needed, E and F. Finally, block G of the third stage of the tree generates two other new partial products to the final adder. This architecture exhibits some irregularities in the layout since it has a complicated interconnection scheme. Hence, the interconnection wires affect the speed and power dissipation of the adder.

3.2.5 Multiplier's Comparison

The basic array multipliers, like the Baugh-Wooley scheme, consume low-power and exhibit relatively good performances. However, their use can be limited to process operands with less than 16-bits (e.g., 8-bits). For operands of 16-bits and over, the modified Booth algorithm reduces the partial product's numbers by half. Therefore, the speed of the multiplier is reduced. Its power dissipation is comparable to the Baugh-Wooley multiplier due to the circuitry overhead in the Booth algorithm. However, circuit techniques can cause this multiplier to have low-power characteristics as will be discussed in this Chapter. The fastest multipliers adopt the Wallace tree with modified Booth encoding. A Wallace tree would lead, in general, to larger power dissipation and area, due to the interconnect wires. Hence, it is not recommended for low-power consumption applications.

3.3 MULTIPLIER ARCHITECTURE AND SIMULATION METHOD

3.3.1 Architecture

The modified Booth multiplier is selected due to its compatibility with low-power operation as discussed in the previous Section. A comparison of different multiplication algorithms presented in [4] has revealed that for the range of 16-32 bit multipliers, the modified Booth algorithm provides a high-performance and lower power dissipation than that of the Wallace or/and Dadda multiplication algorithms. The architecture of a 16x16 bit modified Booth multiplier is shown in Figure 3.9.

On the left hand side are the Booth encoders, one for each partial product. They each have three bits of Y as input (with '0' to the right of the LSB). They are also responsible for decoding and propagating the sign extension logic to the next encoder. The encoders generate five control signals with each signal corresponding to one of $\{-2,-1,0,1,2\}$. The array cells then generate the appropriate bit and add it to the accumulated sum with their internal full adders. For column i, if the control signal is 2x or -2x, the multiplication is performed by selecting X_{i-1}, effectively shifting the partial product to the left. Inversion (-1x or -2x) is handled inside the cells for each bit. In the two's complement format, inverting a number consists of flipping all bits and adding 1. The 'ADD' cell generates the 1 if required. The final stage is a 32-bit adder consisting of two 16-bit adders chained together.

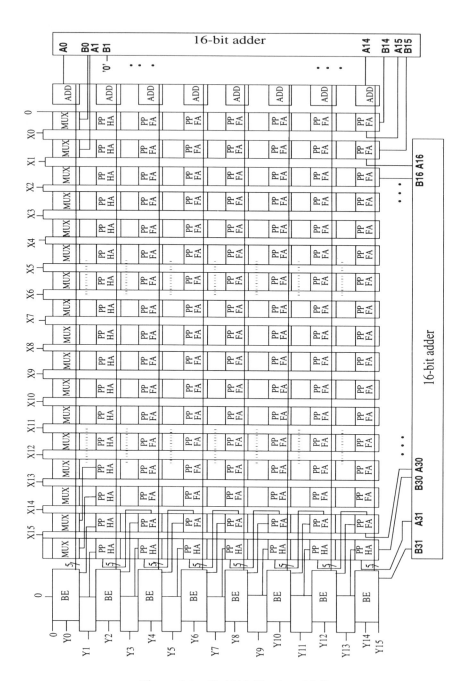

Figure 3.9 16x16 bit Booth multiplier

3.3.2 Simulation Strategy

Simulation of a complex circuit such as a 16x16-bit multiplier is slow and difficult to use in the initial stages of the design process. Fortunately, the modified Booth multiplier architecture (see Figure 3.9), is a regular array of identical cells. It is therefore possible to replace most of the cells by their equivalent input capacitance and to study the performance of only a few of the basic building blocks under appropriate loading conditions. After the initial design is complete, a full simulation is then run to verify operation and obtain more accurate results.

3.3.2.1 Capacitance Estimation

The load on the simulated circuits was a combination of real circuits and the estimated capacitance, C_L (see [5]).

$$C_L = \sum_i C_{g_i} + \sum_m C_{d_m} + \sum_k C_{wire_k} \qquad (3.10)$$

$$C_{g_i} \approx \frac{\epsilon_0 \epsilon_{siO_2}}{t_{ox}} A_{g_i} \qquad (3.11)$$

where A_{g_i} is the area of the gate of transistor i.

$$C_{d_m} = C_{ja}(a_m b_m) + C_{jp}(2a_m + 2b_m) \qquad (3.12)$$

where C_{ja} is the junction capacitance per area, C_{jp} is the periphery capacitance per length, a_m is the width of the diffusion region of transistor m and b_m is the length of the diffusion region of transistor m.

3.3.2.2 Simplified Circuit

Figure 3.10 is a block diagram of the simplified circuit for simulation. The circuit is a simplified version of one row in the multiplier array consisting of the cells under study loaded with equivalent input capacitances. Routing capacitance for long wires was estimated at $0.25pF$. Table 3.4 summarizes the values calculated for the equivalent load capacitances based on the different full adder used in realizing the multiplication array discussed in the next Section.

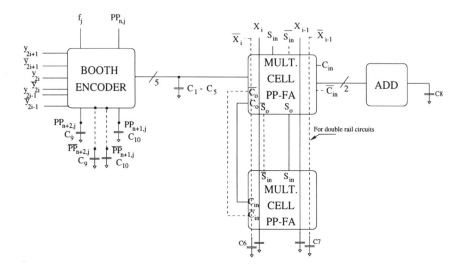

Figure 3.10 Simplified circuit

Table 3.4 Equivalent capacitances in simplified circuit

FA Circuit	Name	Value(pF)
CMOS	C1–C3	0.37
CMOS	C4,C5	0.55
CMOS	C6,C7	0.35
CMOS	C9	0.1
CMOS	C10	0.08
LCPLX	C1-C5	0.86
LCPLX	C6,C7	0.34
LCPLX	C9	0.13
LCPLX	C10	0.07
CPL-TG	C1-C5	0.62
CPL-TG	C6,C7	0.31
CPL-TG	C9	0.12
CPL-TG	C10	0.04
ALL	C8	0.05

Low-Power High-Performance Multipliers 47

In the following Sections, each of the building blocks of the multiplier is discussed and optimization for a 16-bit implementation is presented. Later in Section 3.8 a 6-bit modified Booth is implemented as a test vehicle to test the performance of the designed blocks and to study the tradeoffs of power dissipation and speed of the different design alternatives. The 6-bit implementation is chosen for simplicity, i.e. to reduce the simulation time and circuit complexity.

Since all circuits used to construct the multiplier are CMOS circuits, the simulation used 0.8 μm CMOS (in the BiCMOS process) with the key parameters presented in Table 6.1. The frequency of operation and supply voltages used are 20 MHz and 3.3 V, respectively. A 20 MHz simulation frequency is sufficient due to the lower clock rate the multipliers work at within a DSP application. Also, higher frequencies are no needed to measure the delay (which is independant of the frequency) and power dissipation (which is proportional to the frequency of operation).

3.4 MULTIPLIER CELL

The multiplier cell represents one bit in a partial product and is responsible for:

1. Generating a bit of the correct partial product in response to the signals from the Booth encoder.

2. Adding this bit to the cumulative sum propagated from the row above.

The cell consists of two components, a multiplexer to generate the partial product bit (PP-MUX) and a full or half adder to add this bit with the previous sum (FA or HA). As can be seen in figure 3.9 the first row of the multiplier only needs to generate a partial product and therefore each cell in this row consists only of a PP-MUX circuit. Figure 3.11 shows a block diagram of the multiplier cell.

3.4.1 Full and Half Adder Circuits

The full adder (FA) is the most critical circuit in the multiplier as it ultimately determines the speed and power dissipation of the array.

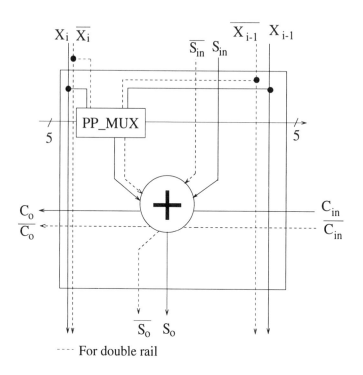

Figure 3.11 Multiplier cell (PP-FA)

The boolean expression for a half adder (HA) is:
$$S_o = A \oplus B \tag{3.13}$$

$$C_o = AB \tag{3.14}$$

and for a full adder:
$$S_o = A \oplus B \oplus C_{in} \tag{3.15}$$

$$C_o = AB'C_{in} + A'BC_{in} + AB \tag{3.16}$$

Low-Power High-Performance Multipliers 49

In order to select the best FA suited for low-power and high-performance applications, a study of existing and novel FA circuits has been done. In total seven CMOS full adder circuits have been studied as shown in Figure 3.12. The circuits are:

1. CMOS: Conventional static CMOS (benchmark)
2. TG-CMOS: Transmission Gate CMOS FA
3. CPL: Complementary Pass Logic FA
4. LCPL1: Latched Complementary Pass Logic (style #1) FA
5. LCPL2: Latched Complementary Pass Logic (style #2) FA
6. DPL: Double Pass-transistor Logic FA
7. CPL-TG: A CPL and TG combination (novel circuit)

The HA circuits are then generated from the optimized FAs by eliminating the circuitry which implements the effect of the input carry.

3.4.1.1 Conventional CMOS FA

The CMOS full adder is used as a benchmark for comparison between all circuits. Small size input inverters ($W_p/W_n = 4/2$) are used to provide the complementary inputs required by the full adder logic since only a single output is provided by the previous stage in the multiplier array. A disadvantage of the CMOS full adder is that the input capacitance can be very high since all the input signals are driving MOS gates. As well, larger PMOS transistors comprise half of the total transistor count, increasing the silicon area. A CMOS FA also suffers from a large delay path compared to other circuits. The optimized CMOS full adder is show in Figure 3.12(a).

3.4.1.2 TG-CMOS FA

The TG-CMOS full adder uses transmission gate logic to realize a complex logic function (ex. XOR) using a minimum number of complementary transistors. Similar to the conventional CMOS, it requires complementary inputs. A TG-CMOS FA exhibits better speed than the conventional one due to the small

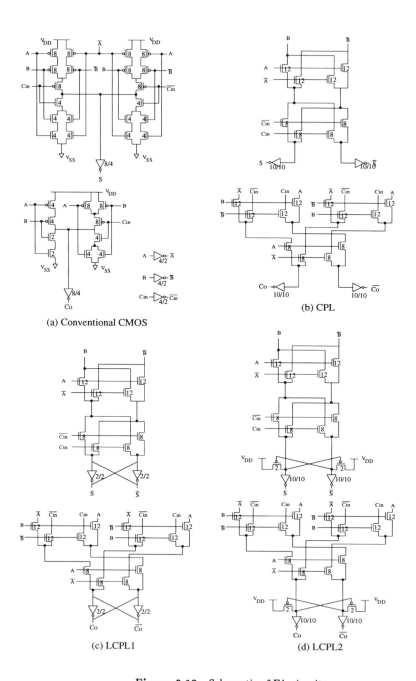

Figure 3.12 Schematic of FA circuits

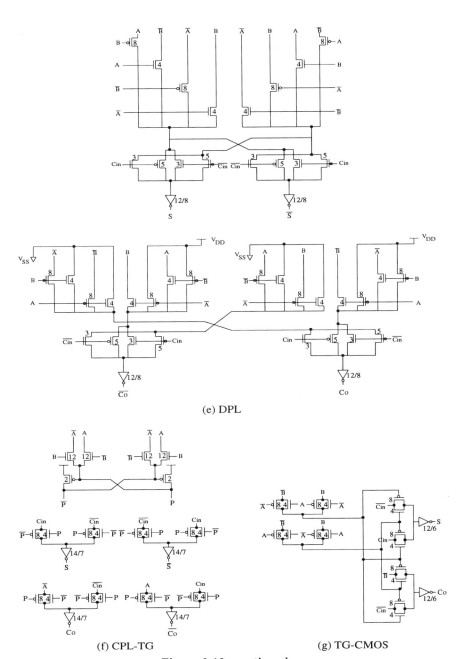

(e) DPL

(f) CPL-TG

(g) TG-CMOS

Figure 3.12: continued.

transistor stack height, at the same power dissipation. The optimized version of the TG-CMOS circuit is shown in Figure 3.12(g)

3.4.1.3 CPL, LCPL1 and LCPL2 FAs

These CPL circuits feature complementary inputs/outputs using NMOS pass-transistor logic with CMOS output inverters. The complementary nature of this style eliminates the need for input inverters. One important feature of CPL circuits is the small stack height and the internal node low-swing which contribute to lowering the power consumption. The pass transistor logic is able to realize complex functions with a minimum number of transistors. The CPL sum circuit consists of two 4-transistor XOR modules in series. The CPL FA suffers from static power consumption due to the low-swing at the gates of the output inverters (Figure 3.12(b)). To lower the power consumption of CPL circuits, two circuit styles are proposed. The output levels are restored with cross-coupled inverters (LCPL1) or with a latch employing minimum size PMOS restoration transistors (LCPL2). The LCPL1 and LCPL2 FAs are shown in Figure 3.12(c) and (d), respectively.

3.4.1.4 DPL FA

The DPL FA is similar in architecture to the CPL counterparts, but it uses complementary transistors to keep full swing operation and reduce the DC power consumption. This eliminates the need for restoration circuitry. One disadvantage of DPL is the large area used. This is due to the presence of PMOS transistors of sizes double that of the NMOS counterparts. The DPL FA is shwon in Figure 3.12(e).

3.4.1.5 CPL-TG

The CPL-TG FA circuit uses CPL XOR logic to generate the signal $P = A \oplus B$. The use of a CPL circuit at the front end reduces the input capacitance due to the small number of transistors. Then, CMOS transmission gate logic is used at the back end to consider the effect of C_{in} and to produce a full swing output. The CPL-TG FA circuit is shown in Figure 3.12(f).

3.4.1.6 FA Simulation Results

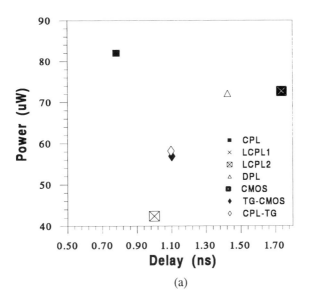

(a)

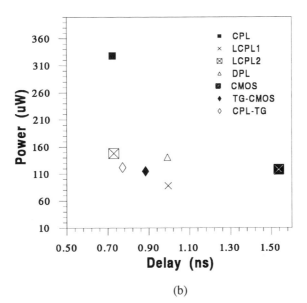

(b)

Figure 3.13 Full adders delay vs power; (a) minimum size (b) optimized for speed

The various full adder circuits were simulated using a multiplier array configuration for proper loading. The optimization process starts with a minimum transistor size design (low-power) and then it optimized for speed. All the delay paths are simulated and the most critical delay is reported.

The simulation results are shown in Figure 3.13(a) for minimum size and Figure 3.13(b) for the optimized version. It is worth noting here the high power dissipation of the CPL circuit due to the static power. The LCPL1 and LCPL2 versions exhibit a lower power consumption than the CPL counterparts due to the elimination of wasted static power. The LCPL1 circuit failed to operate at minimum size transistors due to the inability of the output latch to switch in response to the slow rise output.

Compared to the optimized conventional CMOS FA, all other optimized circuits exhibit better performance. The CPL-TG FA outperforms the conventional CMOS by a speed factor of 2x at the same power dissipation. This translates into an energy saving of 50%. The TG and LCPL2 circuits provide a speed enhancement over conventional circuits of 1.75x and 2.2x, respectively, with energy (power-delay product) savings of 44%. The DPL FA outperform the conventional circuit by a factor of 1.5x in terms of speed, with an energy saving of 25%. This is due to the larger power dissipation than for the conventional circuit.

3.4.1.7 FA Experimental Results

The optimized CPL-TG, LCPL2 and conventional circuits are used to compare the power and delay effects on the multiplier array. A test chip containing the implementation of the three full adders has been fabricated in 0.8 μm CMOS (in BiCMOS) technology through the Canadian Microelectronics Corporation . The photomicrograph of the three circuits are shown in Figure 3.14. The area comparison of the three styles is shown in Table 3.5. The LCPL2 and CPL-TG offer a reduced area compared to that for the conventional circuit.

The layout of the FAs shown in Figure 3.14 consists of a chain of 10 FAs with a separate supply provided to the fifth FA. The separate supply will facilitate the measuring of power dissipation and the 10 stages of FAs is needed to facilitate the delay measurement.

The testing of the FAs are carried out with an input signal at a frequency of 1MHz and a supply voltage of 3.3V. The experimental results for the FAs are shown in Figure 3.15. The input waveform is at the bottom of the figure and

Low-Power High-Performance Multipliers 55

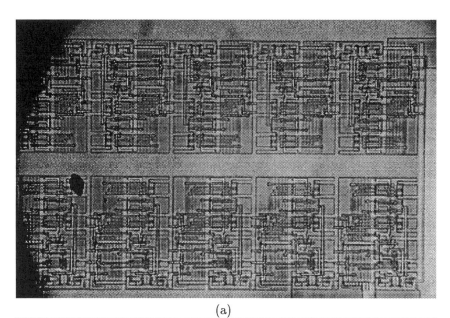

(a)

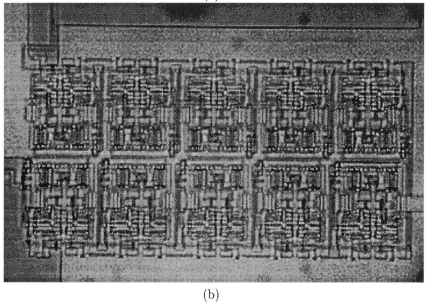

(b)

Figure 3.14 Photomicrograph of FA circuits: (a) conventional, (b) CPL-TG, (c) LCPL2 and (d) Test chip

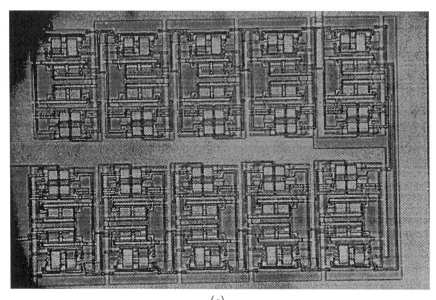

(c)

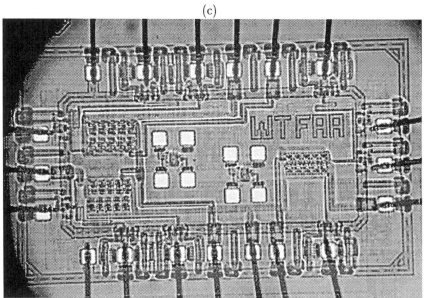

(d)
Figure 3.14: continued.

Low-Power High-Performance Multipliers

Table 3.5 Physical dimensions of full adders

Full Adder	Width (μ)	Height (μ)	Area (μ^2)
CMOS	100	70	7000
CPL-TG	101	57	5757
LCPL2	78	60	4680

the output waveforms of the three FA types are at the top of the figure. Note that the delay of the CMOS-FA is the largest among the three FAs.

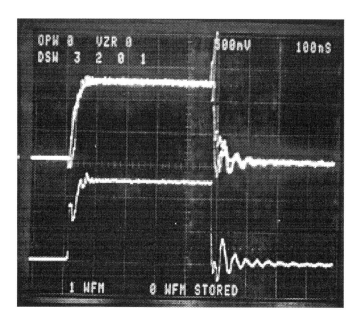

Figure 3.15 FAs experimental verification: input waveform (bottom) and output superimposed waveforms (top).

The measured delay (Td) and the RMS power (Prms) for the conventional CMOS, the CPL-TG, and the LCPL2 FAs are presented in Table 3.6. The measured results agree with the simulated results. The CPL-TG and the LCPL2 FAs provide a speed enhancement of 1.6x and 1.8x, respectively. The little discrepancy in the delay measurement is attributed to the routing capacitance.

Table 3.6 Measured results of optimized full adders

Circuit	Td (ns) simulated	Prms (μW/MHz) simulated	Td (ns) measured	Prms (μW/MHz) measured
CMOS	1.53	6.70	1.47	6.10
CPL-TG	0.88	7.05	0.92	6.60
LCPL2	0.74	7.85	0.79	8.58

The power dissipation of the LCPL2 circuit is the largest among the three as expected, but its power-delay product is less than that of the CMOS. The CPL-TG consumes similar power to the CMOS circuit but with much less delay. The CPL-TG provides the lowest power-delay product.

3.4.2 Multiplexer

The multiplexer (mux) is responsible for generating one bit of the partial product in response to the five control signals from the Booth encoder.

Two designs of the mux are shown in Figure 3.16. One is for single rail circuits (i.e. conventional CMOS) the other is for double rail circuits (i.e. CPL-TG and LCPL2).

Both circuits are optimized for minimum energy consumption. For the case of the single rail circuit, the output buffers are optimized to consume minimum energy as shown in Figure 3.17. The optimized sizing of the double rail mux varies depending on the circuit it is driving. The sizes are indicated in Table 3.7. The delay paths for the mux are equalized to provide equal fall and rise times.

Table 3.7 Optimized transistor sizes for double rail mux (Figure 15(b))

Full Adder	W (μ)
LCPL1	10
LCPL2	10
CPL-TG	6

Low-Power High-Performance Multipliers 59

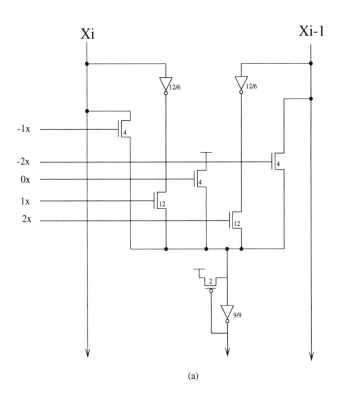

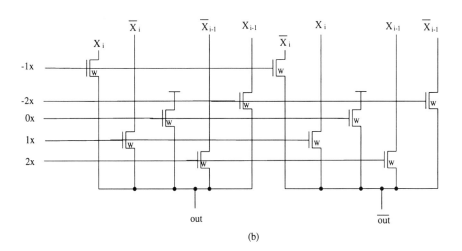

Figure 3.16 Multiplexer circuits: (a) Single rail (b) Double rail

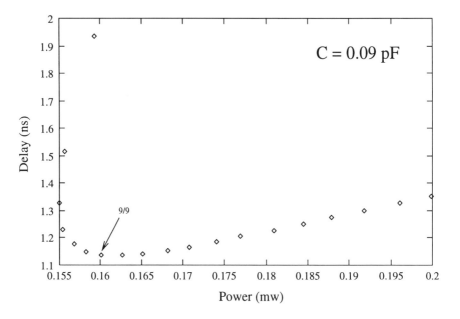

Figure 3.17 Delay vs. Power for single rail mux at various output buffer $\frac{W_p}{W_n}$ sizes.

Low-Power High-Performance Multipliers 61

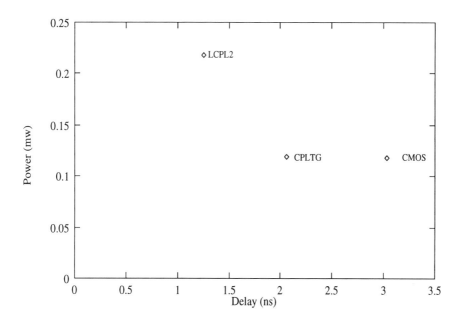

Figure 3.18 Delay vs. Power comparison of multiplier cells

3.4.3 The Multiplier Cell Performance

Three implementations of the multiplier cell are simulated using the simulation method of Figure 3.10. The three implementations are:

1. CMOS = conventional CMOS FA with a single rail mux.

2. LCPL2 = latched CPL FA style #2 plus the double rail mux.

3. CPL-TG = CPL-TG FA plus the double rail mux.

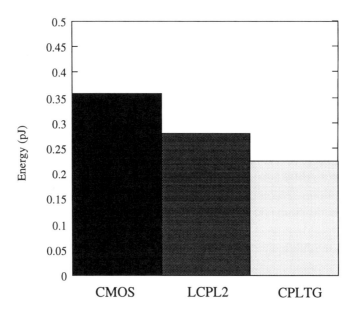

Figure 3.19 Comparison of the delay-power product of multiplier cells

The power vs delay curve for the various circuits is presented in Figure 3.18 . The CMOS version produces low-power at the expense of large delays. The LCPL2 produces a fast cell (about 3x faster than the CMOS cell) but at double the power consumption of its CMOS counterpart. The CPL-TG design provides a 1.5x speed enhancement over the CMOS design at the same power dissipation. A comparison of the energy used by each circuit is shwon in Figure 3.19. The CPL-TG uses the least energy and will be used in the 6-bit multiplier

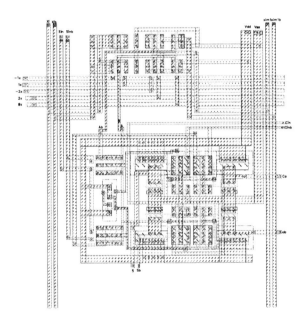

Figure 3.20 Layout of the CPL-TG multiplier cell

implementation. The layout of the CPL-TG multiplier cell is shown in Figure 3.20 .

3.5 BOOTH ENCODER

3.5.1 Logic

The Booth encoder implements the Booth algorithm encoding of three bits of the multiplier Y as well as handling the sign extension logic. It is assumed that the bits $Y_{2i-1}, Y_{2i}, Y_{2i+1}$ and their complements $Y'_{2i-1}, Y'_{2i}, Y'_{2i+1}$ are available from the input register. The logic needed to implement the encoder along with the static CMOS implementation are shown in Figure 3.21.

Each encoder is dedicated to one partial product in the array. Since there is a circuit for each of the five possible generated partial product signals, one and *only* one signal is high during steady state operation. The carry propagate circuits are independent of the partial product circuits and do not share any inputs.

3.5.2 Circuits

The encoder circuits must be able to drive relatively large loads in the order of $0.3pF - 0.6pF$ as they drive 16 or 17 NMOS gates. Two implementations of the encoder have been simulated. The Conventional static CMOS implementation is shown in Figure 3.21 and the CPL implementation is shown in Figure 3.22. The double buffer chain circuit shown in Figure 3.23 is used at the output of the encoder to provide drivability to the large load. Figure 3.23 shows the transient response of the encoder 1X circuitry with and without the buffer chain.

A comparison between the optimized conventional circuit and CPL encoder implementations is shown in Figure 3.24. The CPL implementation provides a speed enhancement of 6% and a power saving of 30%. The low-power characteristics of the CPL encoder and its speed enhancement is very useful for the multiplier design. The layout of the CPL encoder is shown in Figure 3.25.

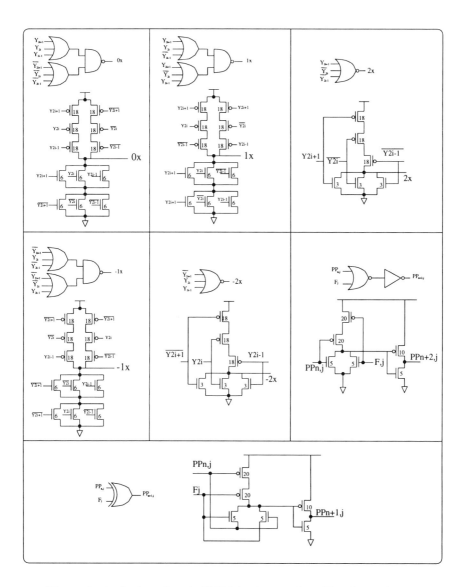

Figure 3.21 Logic and static CMOS implementation of Booth encoder.

66

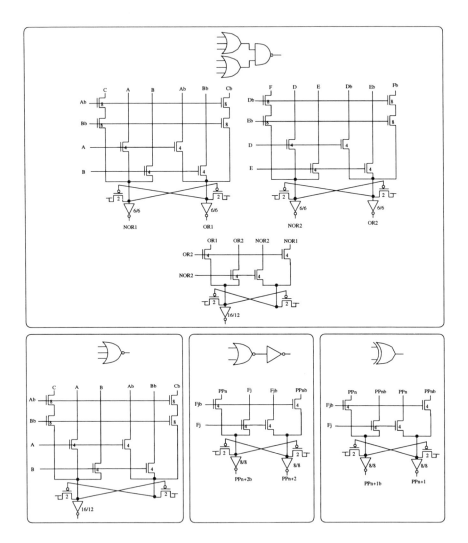

Figure 3.22 CPL implementation of Booth encoder.

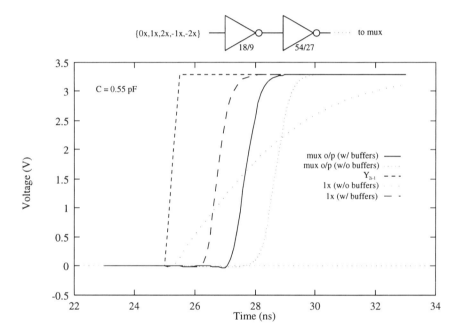

Figure 3.23 Transient response of Booth encoder with and without output buffers

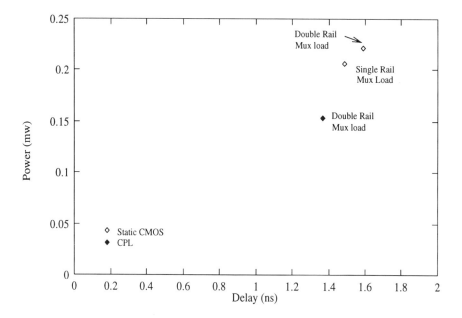

Figure 3.24 A Power and performance comparison of Booth encoder circuits.

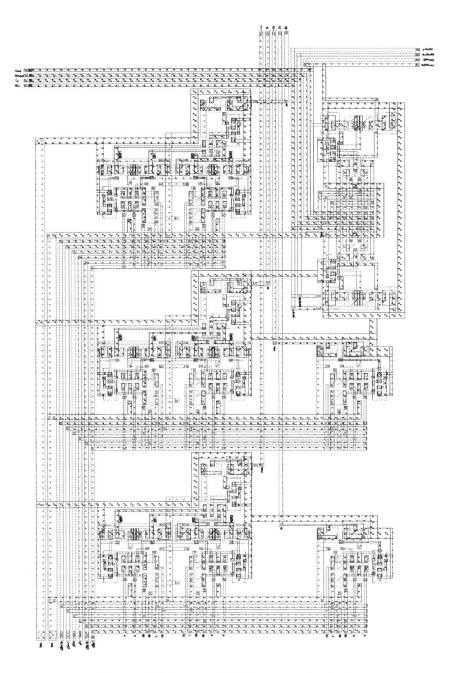

Figure 3.25 Layout of CPL Booth encoder circuit.

Table 3.8 Add cell performance

Delay	0.88 ns
Power	0.07 mW
Energy	0.06 pJ

3.6 ADD CELL

To realize the two's complement of a binary number, first the number is inverted and a logic "1" is added to it. The add circuit generates the "1" to be added if necessary. Figure 3.26 shows the add circuit with optimized transistor sizes. Table 3.8 presents the power and delay of the optimized add cell.

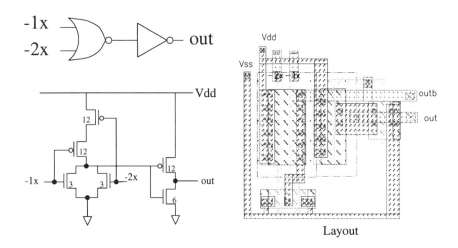

Figure 3.26 Add cell

3.7 CSA

The final addition is performed by a 32-bit condition sum adder with the carry select architecture (CSA) previously presented in Chapter 2. The CPL-like implementation of the CSA is used since it results in a lower power dissipation

Table 3.9 Explanation of the four 6x6 multiplier implementations.

Style	Explanation
CMOS-CMOS	static CMOS encoder + static CMOS FA (fully single rail)
CMOS-LCPL2	static CMOS encoder + LCPL2 FA (array double rail)
CMOS-(CPL-TG)	static CMOS encoder + CPL-TG FA (array double rail)
CPL-(CPL-TG)	latched CPL encoder + CPL-TG FA (fully double rail)

without affecting the adder performance. For further explanation of the CSA adder, refer to Chapter 2.

3.8 6-BIT MULTIPLIER

The circuit architecture of the 6-bit multiplier is described in Figure 3.27. Three circuit style implementations of the 6X6 multiplier were simulated in order to study the effect of the circuit style on the multiplier power and delay. Table 3.9 provide an explanation of the four implementations.

Table 3.10 summarizes the performance of the multipliers. The delay measurement is taken from the critical path in which the slowest signal has to propagate though three rows before it arrives at the CSA.

Similar to the results of the simplified circuit, the CMOS-CMOS implementation dissipates low power but at the expense of a large delay. The CMOS-LCPL2 implementation offers good performance at the expense of doubling the power dissipation. The CMOS-(CPL-TG) on the other hand provides high-performance at approximately the same power dissipation. In fact, the CMOS-(CPL-TG) offers the lowest energy consumption of 9.6 pJ compared to 12.54 pJ and 16.45 pJ for the CMOS-CMOS and the CMOS-LCPL2, respectively.

Since the CMOS-(CPL-TG) implementation offers the best compromise between power and speed, the CPL encoder was combined with the CPL-TG FA to further enhance the multiplier's performance. In Table 3.10, the results of

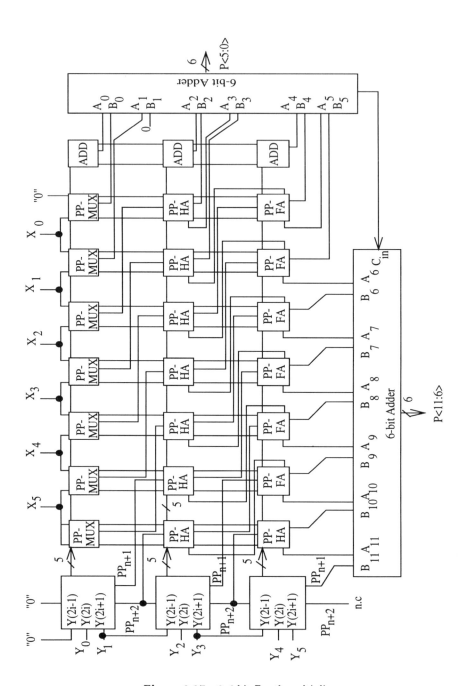

Figure 3.27 6x6 bit Booth multiplier

Table 3.10 6-bit multiplier performance

Name	Critical path	Delay (ns)	Total Power (mW)
CMOS-CMOS	X_5 to P_7	7.25	1.73
CMOS-LCPL2	X_5 to P_7	4.94	3.33
CMOS-(CPL-TG)	X_5 to P_6	5.53	1.74
CPL-(CPL-TG)	X_5 to P_6	4.98	1.42

simulation are shown to be in agreement with the expectation. The CPL-(CPL-TG) provides a speed enhancement of about 10% and a 20% power savings over the CMOS-(CPL-TG) counterpart. This translates into a total energy savings of 26%.

Figure 3.28(a) and (b) shows the effect of the FA array style on the multiplier delay and power, respectively, (using a CMOS encoder). It is clearly shown that the delay and power consumption is attributed to the choice of the FA logic style. The CPL-TG FA provides the best power-delay product, as discussed above.

Figure 3.29(a) and (b) shows the impact of, encoder circuit style on the multiplier delay and power, respectively, (using a CPL-TG array). This comparison reveals that the power dissipation of the CPL encoder is $\frac{1}{2}$ that of the CMOS counterpart. This saving constitutes a 20% saving for the CPL-(CPL-TG) multiplier compared to the CMOS-(CPL-TG) counterpart. The delay of the CPL encoder is reduced by 25% over CMOS counterparts. This translates into a 10% delay reduction of the CPL-(CPL-TG) multiplier over the CMOS-(CPL-TG) counterpart.

3.8.1 Layout of 6-bit multiplier

The multiplier consists of a regular array of cells which must fit together in a compact way. It is therefore necessary to develop a layout strategy before layout is attempted. Figure 3.6 shows the floorplan for an n-bit Booth multiplier.

Figure 3.30 illustrates the proposed layout strategy. The cells are laidout individually and tested. Then they are placed in the upper hierarchy to form the multiplier. The routing of the placed cells is done through the provided channels with metal 1 running vertically and metal 2 running horizontally.

The layout of the multiplier circuit is shown in Figure 3.31. The layout is a direct mapping of the simulation strategy. A test chip of the multiplier is shown in Figure 3.32.

3.9 SUMMARY

Circuit techniques for low-power high-performance multipliers are presented. The modified Booth algorithm is used as a test vehicle. All circuits were optimized for a 16x16 bit multiplier using the 0.8 μm BiCMOS technology.

Simulation of novel and existing full adder circuits revealed that the CPL-TG FA implementation offers an energy saving of 50% over conventional FA. This result is validated by experimental testing of the FAs. Two Booth encoder implementation based on conventional CMOS and CPL were simulated with the former providing a 30% power reduction and 6% speed enhancement.

The dramatic effect of circuit style is demonstrated by simulating a 6x6 multiplier. The combining of a CPL encoder and the CPL-TG FA have resulted in an 18% power saving and a 30% speed enhancement over the conventional CMOS implementation.

The CPL-TG FA and the CPL encoder consume less area than the conventional CMOS FA and CMOS encoder, respectively. This is due to the elemination of the PMOS devices from the circuits.

Low-Power High-Performance Multipliers

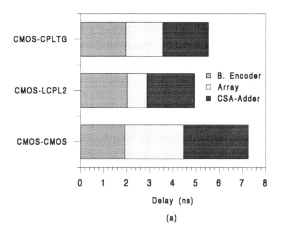

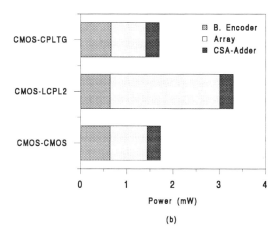

Figure 3.28 Effect of FA style on the multiplier performance; (a) delay and (b) power dissipation

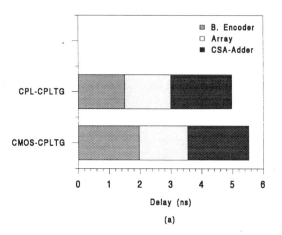

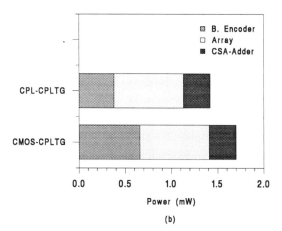

Figure 3.29 Effect of the encoder style on the multiplier performance; (a) delay and (b) power dissipation

Low-Power High-Performance Multipliers 77

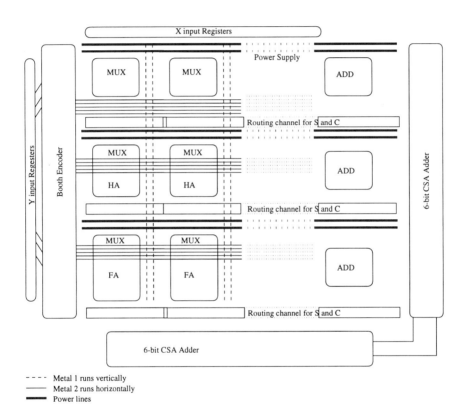

Figure 3.30 Layout strategy

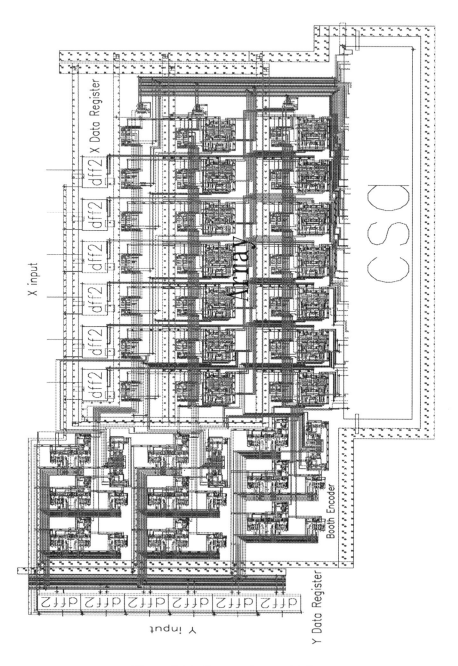

Figure 3.31 Layout of 6x6 multiplier.

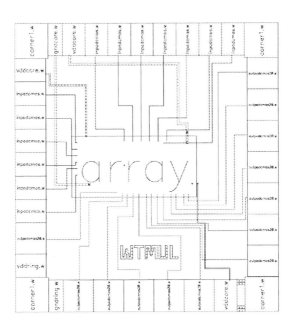

Figure 3.32 Test chip of the multiplier.

REFERENCES

[1] Nomura et al, "A 300-MHz 16-b 0.5 μm BiCMOS Digital Signal Processor Core LSI", IEEE Journal of Solid State Circuits, Vol. 29, No. 3, pp. 290-297.

[2] K. Hwang, "Computer Arithmetic: Principles, Architecture, and Design", John Wiley and Sons, 1979.

[3] J.J.F. Cavanagh, " Computer Science Series: Digital Computer Arithmatic", McGraw-Hill Book Co., 1984.

[4] Ahmad R. Fridi, "Partial Multiplication; A Low-Power Approach for Parallel Multiplier", ECE729 Course Project, Department of Electrical and Computer Engineering, University of Waterloo, April 1994.

[5] N.H.E. Weste and K. Eshraghian, "Principles of CMOS VLSI Design: A systems Perspective", Addison-Wesley Publishing Company, 2nd ED, 1993.

4
LOW-POWER REGISTER FILE

4.1 INTRODUCTION

The advancement in microprocessors and high speed digital signal processors over the past few years is attributed mainly to parallelism and the integration density of VLSI chips. This meant the integration of complete systems on a single chip with many units functioning in parallel to improve the performance and increase the throughput of the system. Parallel operation of these units requires parallel access to the data stored in memory (RAM). To facilitate such an access to a storage memory, a fast and multi-ported register file is used as a buffer between main memory and the functional units. A major drawback of a multi-ported register file is the increase in area needed to accommodate the extra ports.

The use of a register file will increase the system's throughput by simultaneously supplying the stored data to multiple on chip units. It will further enhance the system's performance by reducing the memory access time delay. In [1], a 17ported register file is used to interact with the different units composing the iWarp microprocessor. A typical size of register file used in microprocessors consists of 32 registers each of 32 bits (i.e 32x32 bits).

In this Chapter we introduce a 32x32 bit register file with a single write and double read ports. Although the design of all elements of the register file is explored, the main contribution is in designing a low-power decoder circuit using CPL circuit logic implementation. Section 4.6 discuss the different decoder implementations and Section 4.7 demonstrate the advantage of using the novel decoders in a 32x32-bit register file.

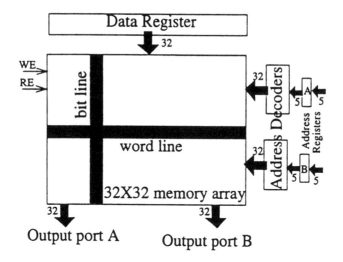

Figure 4.1 Register file architecture.

4.2 ARCHITECTURE AND SIMULATION PROCEDURE

A floor plan of the Register file is shown in Figure 4.1. The register file is composed of a 32x32 memory array, an address decoder, write and read circuits, as well as data and address registers.

The number of ports assigned for this architecture are 3 (2 read ports and 1 write port). Read operations are performed single-endedly; therefore, two read operations can be performed in parallel, one on each bit line. Write operations are performed differentially, and therefore, only one write operation can be performed per memory cycle.

The main building blocks of the register file are designed and simulated using the simplified simulation circuit shown in Figure 4.2. The values of the world line and bit-line capacitances are estimated using 0.8 μm CMOS in BiCMOS technology. In both cases a wiring capacitance of 0.25pF is added to provide a real environment simulation. The values of word and bit line capacitances are indicated on Figure 4.2.

The main objective of this register file design is to reduce the power dissipation while maintaining a high-performance capability. To ensure compliance with

Low-Power Register File 85

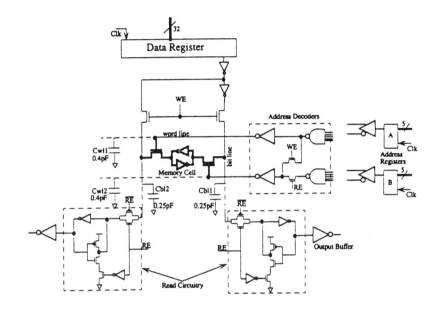

Figure 4.2 Simplified Schematic for simulation of register file.

the main objective, the register file is implemented using static CMOS and CPL circuits. Dynamic circuits were not used. Furthermore, DC and Dynamic power dissipation is minimized by using circuit techniques. This is explained in details in the next Section.

In summary, the register file is designed with the following objectives:

- Minimizing DC and dynamic power dissipation
- Fully static design
- Faster than most CMOS SRAMs (time delay is less than 5 ns)
- Low voltage operation (3.3V)
- implemented using $0.8\mu m$ CMOS in BiCMOS technology

All circuits have been simulated using the $0.8\mu m$ process. A clock frequency of 20 MHz and a supply voltage of 3.3 V have been used.

4.3 MEMORY CELL CIRCUIT

The memory cell used in the register file has a twin port architecture and consists of 6-transistors. The schematic of the memory cell is shown in Figure 4.3(a). The write operation is performed differentially, this ensures fast latching of data and minimizes the power dissipation. The two inverters are of equal size and they exhibit equal rise and fall times.

The simulation results of the memory cell are shown in Figure 4.3(b). The power dissipation is found to be 45.1 μW and the average delay is 0.15 ns. The layout of the memory cell is presented in Figure 4.3(c). It can easily be stacked to form the 32x32 array.

4.4 WRITE CIRCUITRY

The write circuitry consists of a D-type flip flop (D-FF) to store the input data, followed by bit-line buffers to drive the bit-line capacitance. The write enable (WE) signal is connected to the gate of two pass transistors to ensure that only valid data is written to the memory cell. Figure 4.4(a) shows the schematic of the write circuitry. The D-FF circuit is optimized to ensure fast latching of the input data using a master/slave configuration, with the data available at the rising edge of the next clock cycle (i.e. positive edge triggered). To reduce the clock skew a transmission gate is used to equalize the delay between the clk and clkbar signals as shown in Figure 4.4(b).

The use of a single NMOS transistor outputting to the bit line has the following two advantages over the use of a complementary pass gate. First of all, it reduces the voltage swing across a highly capacitive load, which helps in minimizing the dynamic power consumption. Secondly, it eliminates the need for a complementary WE signal. This translates into a reduction in the design complexity and the use of less area. Figure 4.4(c) lists the simulated results for the write circuitry. A total delay of 1.6 ns is needed for the data to be transferred correctly to the bit-lines. The average power dissipation is 0.29 mW per bit-line. The layout of the D-type flip flop used in the write circuit is shown in Figure 4.4(d)

Low-Power Register File

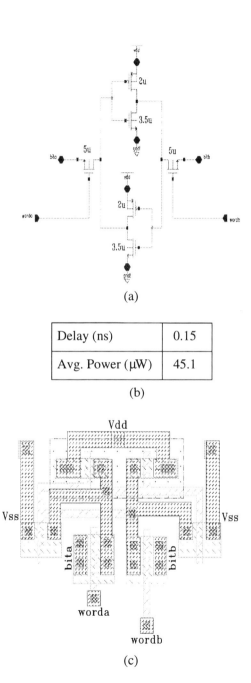

Figure 4.3 Memory cell; (a) schematic, (b) simulation results and (c) layout.

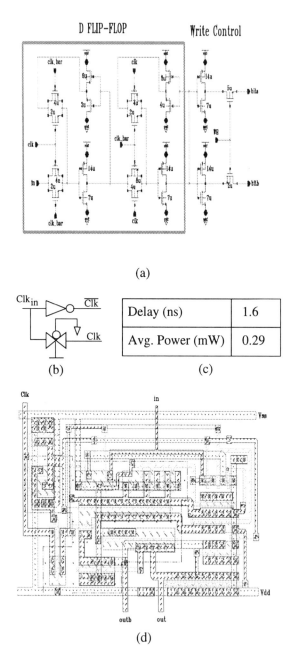

Figure 4.4 Write circuitry; (a) schematic, (b) clock skew reduction (c) simulation results and (d) layout of the D-FF.

4.5 READ CIRCUITRY

The objective of the read operation is to write the data out of the register onto an external bus. A typical load value that a register file will drive is a few pF [2]. A load of 3pF was assumed when designing the read circuit.

There are several factors to take into account when choosing whether to use a sense amplifier or a sense latch. Sense amplifiers are extremely important in large (1Mb,4Mb,16Mb) SRAM chips and are used extensively for that purpose [3–5]. The large load capacitance on the bit lines cause the signal to rise and fall slowly, introducing delays and power dissipation to the circuit. A sense amplifier can detect small variations in the signal. The smaller the memory circuit (10kB or less), the less the load capacitance on the bit lines, and hence, the advantages of using a sense amplifier diminish.

It has been clearly stated in the literature [6] that sense amplifiers consume DC power. Although this DC power consumption can be minimized by using transmission gates or simple pass transistors to cut the DC path when the sense amplifier is not in use, one cannot guarantee its performance for all applications. The success of this method depends on the switching activity of the sense amplifier.

For the reasons discussed above, the design and implementation of a sense amplifier was avoided and a sense latch was implemented instead. The schematic for the sense latch circuit is shown in Figure 4.5(a). The sense latch does not dissipate any DC power and is isolated from the bit-line when the read signal (RE) is low. In this mode the sense latch is acting as a latch storing the read data and making it available to the bus line. When RE is asserted high, the new data is fed through to the read circuitry and latched within when the RE is asserted low.

The simulation results of the read circuitry are shown in Figure 4.4(b), the layout of this circuit is shown in Figure 4.4(c).

4.6 DECODER CIRCUIT

The decoder circuit consists of a 5-bit address register followed by a 5-input AND gate per word. The address register is realized using the D-FF presented in Section 4.3. The decoder circuit has to drive the large load capacitance of

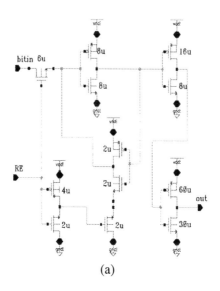

(a)

Delay (ns)	2.2
Avg. Power (mW)	1.01

(b)

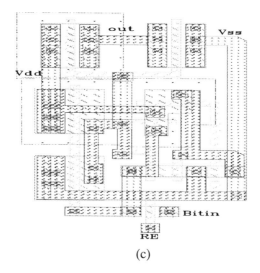

(c)

Figure 4.5 Read circuitry; (a) schematic, (b) simulation results and (c) layout.

Low-Power Register File

the world lines. Hence, it consumes a large portion of the total power and dominates the delay of the register file, and thus posed the main challenge in designing the register file.

Three implementations of the AND/NAND circuit used in realizing the decoder circuit are shown in Figure 4.6. The CMOS static 5-input NAND followed by an inverter is shown in Figure 4.6(a). The CMOS implementation is optimized to provide equal fall and rise times. The CPL implementation of the AND/NAND gate is shown in Figure 4.6(c). A cross coupled latch is used at the output to restore the signal to full swing and to reduce the DC power dissipation. Although the signal and its complement are available at the output, only the AND output is needed to drive the world lines. The CPL exhibits a lower stack height than the CMOS counterpart. The half-CPL implementation shown in Figure 4.6(b) is explored for designs with strict area requirements. A small feedback transistor is used to restore the level and reduce the DC power dissipation. The HCPL exhibits a low input capacitance compared to CMOS and CPL circuits.

Implementation of the AND/NAND gate using a combination of circuits of 2- and 3-input AND/NAND circuits was explored for low power design. It was found that the 5-input configuration will lead to the least delay and a large power savings. This is due to minimizing the switching activity on the internal nodes.

The simulation results for the three circuits are shown in Figure 4.7(a). All circuits have an equal worst-case delay of 1 ns. However, the CPL and HCPL implementations offer power saving of 23% and 18%, respectively. The layout of the HCPL latch is shown in Figure 4.7(b) and is used to implement the decoder circuit. This is due to the fact that initially an HCPL circuit is used based on the need for one signal to drive the world line. The CPL simulation was explored after the chip submission.

4.7 32X32-BIT REGISTER FILE

A 32x32 register file was simulated based on the simulation method presented in Figure 4.2. The waveforms for the write and read operations are shown in Figure 4.8. The simulated delay and average power dissipation of the 32x32 bit register file are presented in Table 4.1. The CPL and HCPL implementations

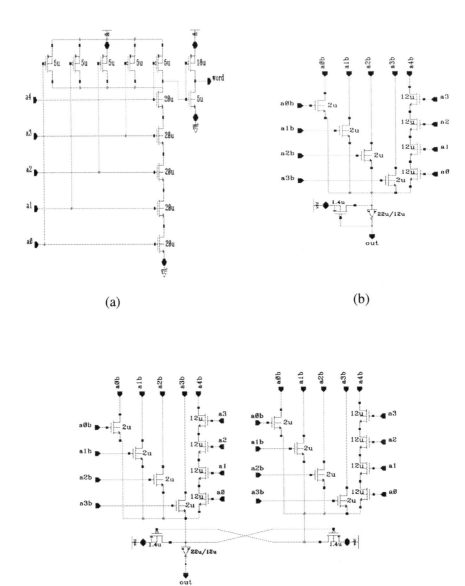

Figure 4.6 AND/NAND circuits; (a) conventional CMOS, (b) HCPL and (c) CPL.

Low-Power Register File 93

	CMOS	HCPL	CPL
Delay (ns)	1.01	1.02	1.03
Avg. Power (mW)	2.6	2.13	2.01

(a)

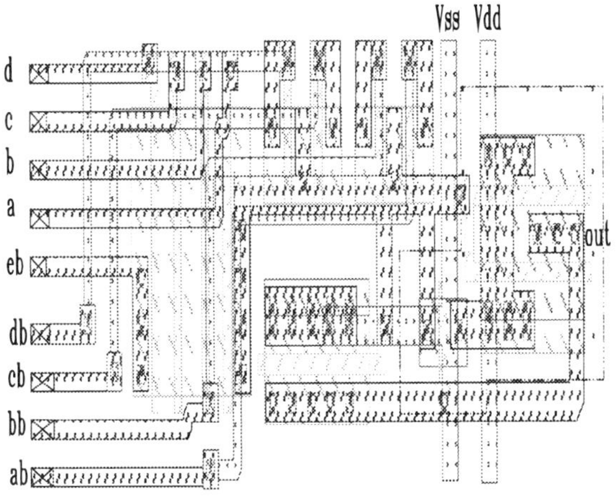

(b)

Figure 4.7 (a) Simulation results, (b) Layout of HCPL circuit.

Table 4.1 Delay and power consumption of the register file.

Decoder Used	Write Delay (ns)	Read Delay (ns)	Average Power (mW/MHz)
CMOS	1.6	2.25	4.84
HCPL	1.6	2.25	4.09
CPL	1.6	2.25	3.89

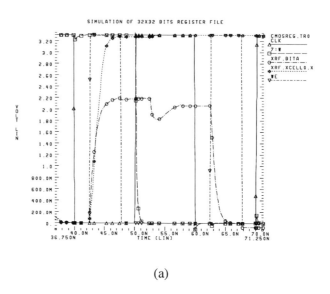

(a)

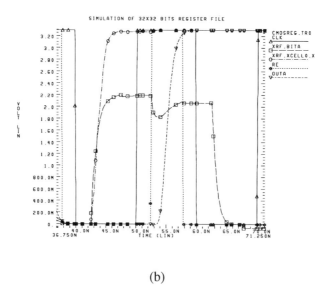

(b)

Figure 4.8 Register file waveforms; (a) write operation, (b) read operation.

Low-Power Register File

offer a power reduction of 20% and 16%, respectively, over the conventional implementation.

A test chip of the 32x32 register file is shown in Figure 4.9. The chip layout follows closely the floor plan presented in Figure 4.1. The decoder was implemented using the HCPL circuit as stated in the previous Section.

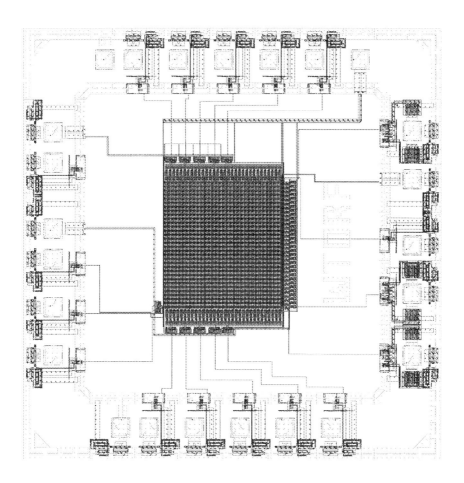

Figure 4.9 Test chip of the 32x32 bit register file.

4.8 SUMMARY

A 32x32 bit register file has been designed and simulated. Circuit design methodology used to reduce power consumption was presented. Three implementations of the register file using conventional CMOS, CPL and HCPL decoder circuits are simulated. The CPL and HCPL circuits provide power savings of 20% and 16%, respectively, compared to the conventional counterpart.

REFERENCES

[1] Jolly, Richard D., "A 9-ns, 1.4-Gigabyte/s, 17-Ported CMOS Register File", IEEE Journal of Solid-State Circuits, Oct. 1991.

[2] S. H. K. Embabi, A. Bellaouar and M. I. Elmasry, "BiCMOS Digital Integrated Circuit Design", Kluwer Academics Pub., MA, 1993.

[3] Ishibashi, K. and et al, "A 12.5-ns 16-Mb CMOS SRAM with Common-Centroid-Geometry-Layout Sense Amplifiers", IEEE Journal of Solid-State Circuits, Apr. 1994.

[4] Seki, T. and et al, "A 6-ns 1MB CMOS SRAM with Latched Sense Amplifier", IEEE Journal of Solid-State Circuits, Apr. 1993.

[5] Seno, K. and et al, "A 9-ns 16-Mb CMOS SRAM with Offset- Compensated Current Sense Amplifier", IEEE Journal of Solid-State Circuits, Nov. 1993.

[6] N.H.E. Weste and K. Eshraghian, "Principles of CMOS VLSI Design: A systems Perspective", Addison-Wesley Publishing Company, 2nd ED, 1993.

5
LOW-POWER EMBEDDED BICMOS/ECL SRAMS

5.1 INTRODUCTION

As was mentioned in chapter 1, one of the most successful applications of BiCMOS technology was in the design of SRAMs [1]. By combining the high-density and low power dissipation of CMOS with the extremely fast speed of bipolar ECL techniques, BiCMOS SRAMs achieved speeds very close to that of bipolar SRAMs at power levels close to that of CMOS SRAMs [2]. BiCMOS ECL SRAMs were usually designed to operate at the maximum power level allowed by still or forced-air cooled plastic packages (600-1000 mW) [3] for maximum speed. With each new generation, the speed was enhanced by scaling the technology and/or using novel circuit techniques and architectures. Hence, the speed was slightly increased, keeping the power consumption at a relatively constant level. This was true for both CMOS and BiCMOS SRAMs. Figure 5.1 shows a comparison of the speed (access time) and power consumption of the latest generations of CMOS and BiCMOS ECL SRAMs. For each generation, the minimum feature size (i.e L_eff) was approximately the same for both types of SRAMs. This figure shows that the speed of BiCMOS SRAMs increased steadily with each generation while the power remained almost constant. This, however, meant that with each new generation, less power was allocated to the ECL I/Os (which increased by two I/Os with each generation). Hence, the speed advantage of these ECL I/Os decreased with each new generation. Coupled with the fact that it is very difficult to implement automatic address transition detection (ATD) techniques with the ECL address inputs (to reduce the overall power consumption), the development of truly asynchronous 16Mb+ ECL SRAMs was very much hindered. Embedded BiCMOS SRAMs face the same power dilemma as stand-alone SRAMs since the power budget allocated for these SRAMs are usually very small.

In the next section the different options for SRAMs front-end organization are studied and the optimum one is identified. In section 3 of this chapter, several new circuit techniques that reduce the power of the ECL address input buffers, Wired-OR (W-OR) pre-decoders, and level-translators, while enhancing their speed, are presented. The power saving advantage of these circuits is demonstrated and compared to conventional techniques. A novel column activation/sensing technique that reduces the DC power of the activated memory block is presented in sections 4 and 5. In section 4, a new self-resetting word-line decoding and driving technique is presented. This new word-line decoder and driver (WLDD) produces a finite pulse that activates the selected word-line for a short period of time. Since this circuit only activates the WL for a short time, the bit-lines signal has to be detected by a latched-sense amplifier. A novel unclocked latched-sense amplifier is presented in section 5. The performance of this new column activation/sensing technique is evaluated and compared to conventional techniques in terms of speed, power, and area.

For the work presented in this chapter, a high-performance, non-complementary (0.35 μm, 3.3V) BiCMOS technology parameters have been used in the simulations. The key device parameters of this technology are shown in Table 5.1.

β	I_K	τ_F	R_B	R_C	C_{eb}	C_{bc}	C_{cs}
90	2 mA	7 pS	520	65	7 fF	9.4 fF	36 fF

(a) The Bipolar (1x) parameters

	V_{th}	T_{ox} (nm)	I_{DSmax} (mA/U)	$C_{S/D}$ (fF/U^2)
NMOS	0.55 V	9.0	0.475 *	0.1
PMOS	-0.55 V	9.0	0.230 *	0.1

* For VGS= 3.3 V

(b) The MOS parameters

Table 5.1 The key device parameters of the (0.35 μm, 3.3V) BiCMOS technology.

Low-Power Embedded BiCMOS/ECL SRAMs

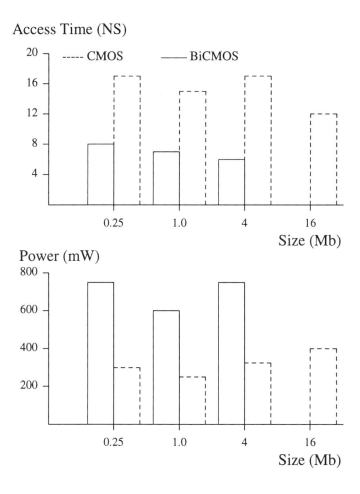

Figure 5.1 Access time and power of the different generations of CMOS/BiCMOS SRAMs.

5.2 16 MB⁺ SRAMS FRONT-END OPTIMIZATION

Typically, the front-end circuitry of a BiCMOS ECL SRAM consumes 40-60% of the total chip power. Hence, any reduction of the power consumption of these circuits would amount to a considerable saving in the overall SRAM's power. A BiCMOS ECL SRAM front-end includes the input address buffers, pre-decoders, and ECL-to-CMOS level-translators.

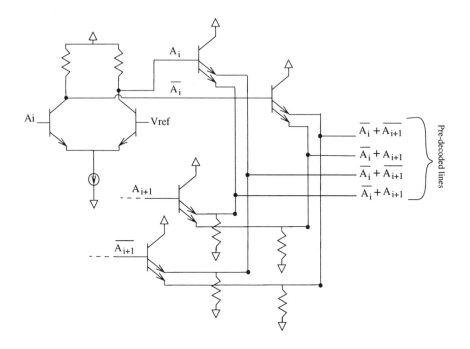

Figure 5.2 The ECL address input buffer and W-OR pre-decoder.

Recent generations of fast BiCMOS ECL SRAMs usually employed a W-OR pre-decoding stage similar to the one shown in Figure 5.2 as a mean of speeding up the decoding path at no extra power [4]. To further enhance the speed at the expense of increased power, the ECL-to-CMOS level translators were moved to the memory blocks. As a result, the pre-decoded signals were ECL levels rather than CMOS levels [4]. While this technique was successful for previous generations of BiCMOS SRAMs, the power limitations associated with 16 Mb (and beyond) SRAMs make this technique slower than the conventional techniques. This placed the level-translators right after the W-OR pre-decoders.

To demonstrate the above fact, several simulations were performed to characterize and evaluate the different SRAMs' front-end options. The different options are : CMOS inputs (an all CMOS front-end), ECL inputs and pre-decoding with level-translation at the memory blocks, or ECL inputs and pre-decoding followed immediately by the level-translators. The difference between the two latter options is the type of pre-decoded signals; should they be ECL levels or

BiNMOS (CMOS) levels. The results are shown in Figures 6.3-6.6. For these results, the ECL buffers were driven by an ECL buffer of a fixed size, and the BiNMOS buffers were driven by a CMOS inverter of a fixed size.

Figure 5.3 shows the power of ECL and BiNMOS buffers at 100 MHz versus the load capacitance for a fixed delay (450 pS). Each buffer was designed to achieve the specified delay at the minimum power for each value of the load capacitance CL. As the load increases, the power difference (ECL power minus BiNMOS power) becomes greater. Hence at higher SRAMs capacity, as the pre-decoded lines get longer, the BiNMOS buffer becomes more power-efficient than the ECL buffer.

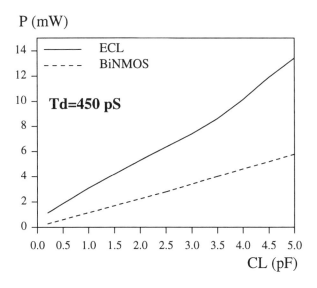

Figure 5.3 Power vs. CL @ 100 MHz for the ECL and BiNMOS buffers for a fixed delay of 450 pS.

It might be argued that at certain value(s) of delay the ECL buffer could become more power-efficient. Figure 5.4 shows the power versus delay for the two buffers for two values of CL. This figure shows that for a larger buffer delay and small CL, the power consumption of the two buffers becomes closer. However, as CL increases the power difference at high delays also increases. Also, operating the ECL buffer at such low speed defies the purpose of using ECL. Another important point to consider is that, for these simulations

(Figure 5.3 and 6.4), a 100% switching activity was assumed. This is a very biased condition against the BiNMOS buffer, since all its power is dynamic (i.e. $P_{BiNMOS} = 0.5 \times Switching\ Activity \times CL \times Freq \times VDD^2$) and only few pre-decoded lines switch with every new memory access. So the actual overall power efficiency of the BiNMOS pre-decoder buffers is much higher than the ECL ones, which mostly consumes DC power.

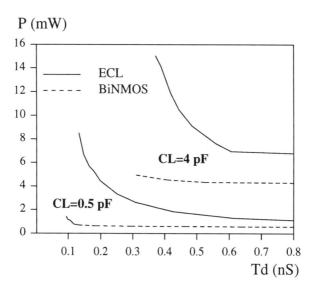

Figure 5.4 Power versus delay for the ECL and BiNMOS buffers for two values of CL @ 100 MHz.

The speed-up of ECL buffers to CMOS and BiNMOS buffers versus the power ratio for several values of CL is shown in Figures 6.5 and 6.6, respectively. The speed-up and power ratio are defined as :

$$Speed - Up = \frac{CMOS\ buffer\ (or\ BiNMOS)\ Delay}{ECL\ buffer\ Delay} \quad (5.1)$$

$$Power - ratio = \frac{ECL\ buffer\ power}{CMOS\ buffer\ (or\ BiNMOS)\ power} \quad (5.2)$$

The CMOS buffer was a multi-stage inverter chain with a fixed size first stage (50 fF input capacitance). The number of stages and their sizing (tapering) were optimized for each value of CL. The BiNMOS buffer was of a fixed size (also 50 fF input capacitance) for all values of CL. The small vertical bars on the different curves represent the maximum power limit of the ECL buffer. This limit is reached when the power of the ECL buffer reaches 20 mW (which translates to about 500 mW total ECL buffers' power for a 16 Mb SRAM). Figure 5.5 shows that the ECL buffer achieves better speed-ups over the CMOS buffer as the load decreases. This, and the fact that ECL signals are better suited for high-speed inter-chip communications (easy to terminate and less supply noise), justify having ECL input buffers rather than CMOS ones. Figure 5.6 shows that for the ECL/BiNMOS speed-up, the trend is the opposite. As CL increases, the speed-up increases. However, due to the ECL power limit, the feasible speed-up does not even reach 0.9 (i.e. the ECL buffer can not be operated even as fast as the BiNMOS one). These two opposite trends of ECL/CMOS and ECL/BiNMOS speed-ups are due to the fact that the CMOS buffer was optimized for each CL, while the BiNMOS buffer was of a fixed size. The BiNMOS could be optimized at high loads by adding a CMOS pre-driver and increasing its size. This would make the two trends similar, and make the ECL/BiNMOS speed-up even smaller at high loads.

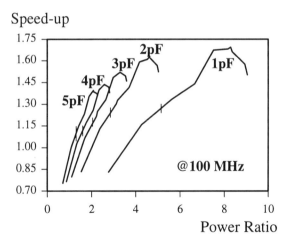

Figure 5.5 Speed-up of ECL over CMOS versus the power ratio for several loads. The vertical bars represent the maximum power limit for the ECL buffer.

Hence, the results shown above illustrate that an optimum front-end combination of a 16 MB[+] SRAM access path would consist of an ECL input buffer and

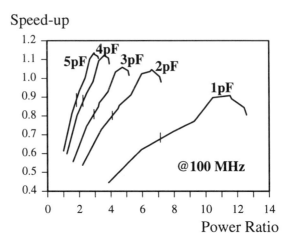

Figure 5.6 Speed-up of ECL over BiNMOS versus the power ratio for several loads. The vertical bars represent the maximum power limit for the ECL buffer.

W-OR pre-decoding. These are immediately followed by level-translation and BiNMOS pre-decoder drivers, respectively.

5.3 THE NOVEL W-ORING AND LEVEL-TRANSLATION CIRCUITS

Conventional W-OR pre-decoders (Figure 5.2) use resistors to pull-down the output. This circuit draws a large power, especially when the output is high. An examination of the four outputs of the pre-decoder shows that for any combination of the two inputs A_i and A_{i+1}, three of the four W-OR outputs would be high as shown in Table 5.2. This means that the conventional W-OR pre-decoders will always have a very high stand-by power, no matter what the input address is.

Recently, several active-pull-down techniques that reduce the power of emitter-followers while maintaining or increasing their speeds were reported [5]- [9]. However, these circuits either (1) can not be W-ORed at all [7], (2) require

A_i	A_{i+1}	$A_i + A_{i+1}$	$A_i + \overline{A_{i+1}}$	$\overline{A_i} + A_{i+1}$	$\overline{A_i} + \overline{A_{i+1}}$
L	L	L	H	H	H
L	H	H	L	H	H
H	L	H	H	L	H
H	H	H	H	H	L

L=Low   H=High

Table 5.2 The truth table of the four outputs of the W-OR pre-decoder.

special devices such as PNPs [5, 6], (3) are very difficult to design properly [8], or (4) they lose their power advantage if they are W-ORed [5, 6, 9].

Three W-OR pre-decoder and level-translator combinations are shown in Figure 5.7. A_i, A_{i+1}, $\overline{A_i}$, and $\overline{A_{i+1}}$ are the outputs of the CML input buffers (which is not shown in that figure). The CMOS inverters at the outputs of the three combinations are used to ensure full-signal restoration and as pre-drivers for the BiNMOS pre-decoded line-drivers. A conventional combination that consists of a resistor-type W-OR and a PMOS common-source level-translator with NMOS current mirror load is shown in Figure 5.7(a). For this combination, the level-translator consumes less power when the output of the W-OR is high. However, it still suffers from the high power problem of the W-OR section.

In the novel combination of Figure 5.7(b), a diode was added to the W-OR section and only one common-source MOS is used for the level-translator section. Since the level-translator is placed right after the W-OR, there is no problem in having two outputs from the W-OR section (i.e. no need to drive long pre-decoded lines). This type of W-OR would not be practical if the level-translators were placed at the memory blocks. The level-translator part of this combination consumes far less power than the conventional technique for both levels of the output of the W-OR output. Its W-OR section, however, still consumes a relatively high power when its output is high.

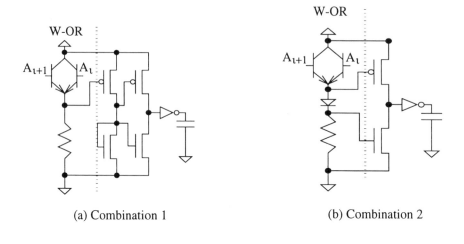

(a) Combination 1   (b) Combination 2

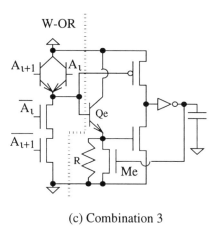

(c) Combination 3

Figure 5.7 The three W-OR pre-decoder and level-translator combinations.

In the other novel combination, Figure 5.7(c), the pull-down circuitry of the W-OR is implemented using NMOS devices that are controlled by the outputs of the input buffers instead of simple resistors. This pull-down circuitry consumes the least amount of power when the output is high (e.g. when both $\overline{A_i}$ and $\overline{A_{i+1}}$ are LOW). The level-translation section is again a single common-source MOS. The level-shifting is achieved by the bipolar Qe and the feedback-controlled

NMOS device Me. The resistor R is set to a very large value and is used to maintain the V_{BE} drop of Qe when the output is low and Me is off. Hence, very small current is drawn by R when Me is off. This level-translator consumes more power than the one used in combination 2 due to the way the level-shifting is achieved.

The results of power optimization for the three combinations of W-OR pre-decoder and level-translator are shown in Figure 5.8. The average power of the W-OR pre-decoder parts (including the power of the CML input buffer) of the three combinations was varied and the level-translator part was sized accordingly to obtain a total delay of 600 pS (with approximately equal rise and fall times). Hence, the power and speed were traded between the two parts of each combination such that the total delay is 600 pS. As shown in Figure 5.8, the power of the W-OR section of combinations 1 and 2 could not be reduced below 9 mW. The conventional combination had an optimum power of 15.6 mW. The new circuits, combinations 2 and 3, had optimum power values of 11.7 mW and 10.1 mW, respectively. This means that, although the level-translator of combination 3 consumes more power than that of combination 2, the power of its W-OR section can be significantly reduced below that of combination 2. This yielded the lowest optimum power.

The above power savings translate to 10-15% and 15-21% total chip power savings, for combinations 2 and 3, respectively. Although combination 2 consumes more power than combination 3, it has smaller area and can be used when area constraints can not be met by combination 3.

5.4 THE NOVEL SELF-RESETTING WL DECODER AND DRIVER

A major source of DC power in ECL BiCMOS SRAMs besides the ECL I/Os, is the memory block(s) that is activated during the memory access. All the columns in a selected block (or sub-block) consume DC power as long as one of that block's word-lines (WL) is activated. Techniques such as array division, divided word-line (DWL) [10], and hierarchical word decoding (HWD) [11], which are used to reduce the decoding delay, also reduce the block size and hence the DC currents in activated memory blocks. However, there is a limit imposed on the amount of array division by the area and delay constraints [10, 11]. For a 16 Mb SRAM that is divided into 64 blocks, the number of columns per block would be 256. This value is still high enough to contribute a

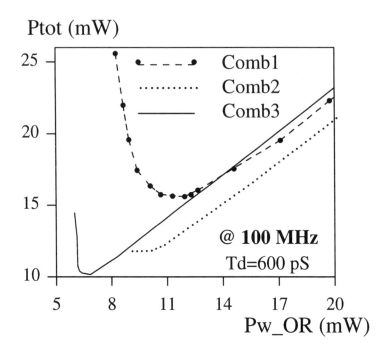

Figure 5.8 The total power versus the power of the W-OR pre-decoder for the three combinations of Figure 5.7.

significant DC current during activation. If each column in the activated block draws a 200 μA current, the total active block current would be more than 50 mA. Hence, the DC power consumed by the activated block would be about 15-25% of the total active power.

Typically, a selected word-line would be activated (asserted high) for a few nanoseconds until the read operation is finished. The actual time required to establish the bit-line signal and to sense it is much less than that. Hence, if a latched sense-amplifier is used, the word-line (and thus all the columns in the selected block) need not be activated for more than the time it takes the latched sense-amplifier to latch-in the data on the bit-lines.

A scheme that reduces the word-line activation time of a 0.5 Mb SRAM was reported in [12]. This scheme was intended to reduce the cycle time and pipeline the memory access operation. It utilizes self-resetting blocks for clock generation and global X/Y/Z-decoder lines. This requires a very tight timing control over these self-resetting blocks to ensure the correct operation of the SRAM. For a 16 Mb+ SRAM this is even more difficult to achieve and would mean that the output pulses of the self-resetting blocks have to be made long enough to ensure proper operation. Thus, the power savings are reduced. Furthermore, pulsing all the global decoded lines with their huge capacitances on average would increase the dynamic power dissipation.

A new localized self-resetting WLDD circuit technique that only activates the selected WL for a short time was developed . With this technique, all the global decoder signals remain static (non-pulsed or clocked) and no timing pulses are required. The circuit diagram of the new WLDD circuit along with the equivalent logic circuit is shown in Figure 5.9. The same circuit can be realized using pure CMOS. Although the circuit appears to be complicated and large, in reality it is not. The devices Mp1 through Mp4, Mn4, and the devices in the inverter driving Mn6 have minimum (or close to minimum) widths. As will be shown later, these devices need not be large for the correct operation of the circuit. Additionally, the two devices Mp6 and Q are merged together to form what is called a BiPMOS device. It has been shown in chapter 4 and in [13] that such devices can achieve fast full swing operation when turned on and then off during output pull-up transition.

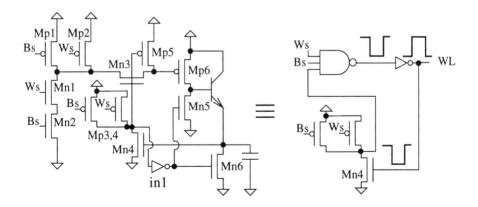

Figure 5.9 The new word-line decoder and driver (WLDD) circuit. The equivalent logic circuit is also shown.

5.4.1 Circuit Operation

Referring to Figure 5.9, the circuit operates as follows: initially the output is low and Mn4, Mn5, Mn6, Mp5, Mp6 are off, and Mn3 is on. The gate voltage of Mp6 will be $\geq V_{DD} - V_{tp}$ (V_{tp} is the PMOS threshold voltage). When both inputs BS and $\overline{\text{WS}}$ (block select and word select) become high, the gate of Mp6 is discharged through the NMOS chain Mn1-3, and Mp6 turns on and starts charging the base of Q. Q then turns on and charges the output node. When the output voltage reaches a certain value, the feedback transistor Mn4 turns on and discharges the gates of Mn3 and Mp5. This turns Mp6 off, causing it to inject half of its channel charge into the base of the merged bipolar Q. This makes Q conduct further making the output reach the full swing. Then the CMOS inverter in1 turns Mn6 on which discharges the output node back to ground.

The width of the output pulse can be controlled by the sizing of Mn4 and the inverter in1. Figure 5.10 shows the output of the self-resetting WLDD for several widths of Mn4 and in1. This figure shows that when both Bs and Ws become high, the new WLDD produces a pulse with a finite width that is proportional to the feedback speed. The pulse width becomes smaller for faster feedback. This figure also shows that there is a slight loss of output swing when the feedback is made very fast (large Mn4 and/or in1). The loss in swing, however, is still insignificant for a pulse width of 600 pS.

5.4.2 Performance Comparisons

The delay of the new WLDD circuit was compared to that of a conventional BiNMOS WLDD technique that is usually used in BiCMOS SRAMs. The BiNMOS WLDD is made of a two-input CMOS NAND gate (with Bs and Ws as inputs), followed by a BiNMOS driver. Both WLDD circuits were designed to have equal input capacitance and area. Figure 5.11 shows the rise delays (from 50% of the input to 50% of the rising output) of both circuits. The delay of the new WLDD is slightly less than that of the conventional one (by 12% at 1 pF load). Furthermore, since the WL is activated only for a short time, larger bit-line loads can be used, enhancing both the read and write speeds.

Although the speed improvement of the new WLDD is not very large, the power saving is very significant. If a WL is activated for 0.6 nS by the new WLDD (compared to 5 nS by a conventional WLDD), the power consumed in

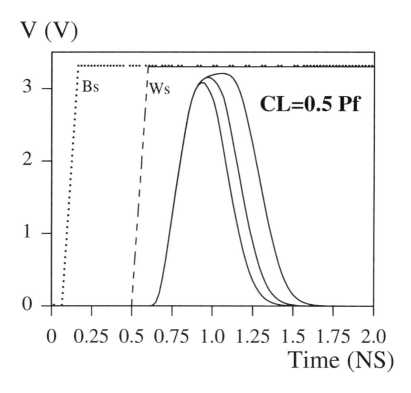

Figure 5.10 The output waveform of the new WLDD for different speeds of feedback.

the selected block would be reduced by more than 8 times. This translates to a 13-22% total chip power saving.

5.5 THE NOVEL LATCHED SENSE-AMPLIFIER

Due to the temporary activation of the bit-lines, a column sensing scheme was required that could sense the bit-lines data and store it quickly before it diminishes. One option is to insert a latch between the local sense-amplifier (or the main sense-amplifier) and the output driver. Such a scheme would have several drawbacks: 1) the WL activation pulse would have to be long enough for the delays through the sense-amplifier(s) and the latch, hence reducing the

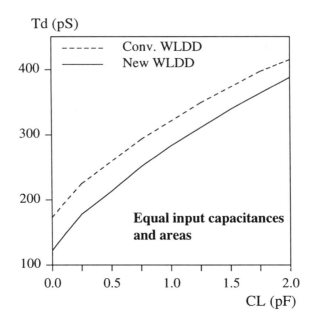

Figure 5.11 The rise delays of the conventional WLDD circuit (CMOS NAND + BiNMOS driver) and the new WLDD circuit versus the load capacitance.

power advantage of the self-resetting WLDD scheme, and 2) a synchronization pulse for the latch would have to be generated and accurately synchronized with the self-resetting WLDD pulse, something very difficult to efficiently achieve in a 16 Mb+ SRAM.

A new unclocked latched-ECL sense-amplifier was especially developed to sense the temporary activated bit-lines data. The new latched sense-amplifier circuit diagram is shown in Figure 5.12. It consists of an emitter-follower input stage followed by an ECL amplifier with latched outputs. The inputs to the emitter-followers come from the local bit-lines Bit and $\overline{Bit}$ (which in turn would be connected to several columns). The signal SA is a static signal that could be the Bs signal or another output signal of the Z-decoder (sub-block decoder).

This sense-amplifier works as follows: when the sense-amplifier is activated by the SA signal, and the bit-lines are still at their pre-charged state (i.e. @

$$\text{Ise R2} = \frac{1}{2} \text{Iss R1}$$

$$\therefore \text{Ise} = \frac{1}{2} \frac{R1}{R2} \text{Iss} \quad \text{or} \quad R2 = \frac{1}{2} \frac{\text{Iss}}{\text{Ise}} R1$$

$$a = \frac{\text{Iss}}{\text{Ise}}$$

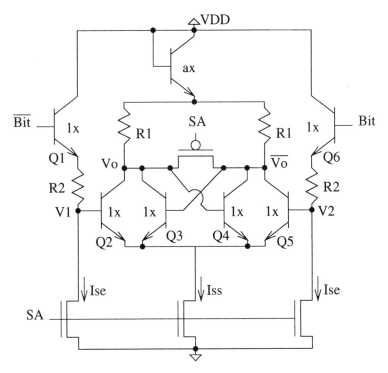

Figure 5.12 The novel latched-ECL sense-amplifier.

VDD), the outputs of the sense-amplifier are discharged from their initial level ($\geq VDD - V_{BE}$) to a level given by the following equation :

$$Vo(t=0^+) = \overline{Vo}(t=0^+) = VDD - V_{BE} - \frac{I_{ss}}{2} \cdot R1 \qquad (5.3)$$

where I_{ss} is the ECL amplifier tail current (Figure 5.12). The tail tail current in this state is equally divided between the two branches of the ECL amplifier. Also, the outputs V1 and V2 of the input level-shifting emitter-followers (at the bases of Q2 and Q5) would be :-

$$V1 = V2 = VDD - V_{BE} - I_{se} \cdot R2 \qquad (5.4)$$

which can be expressed as :

$$I_{se} = 0.5\ I_{ss}\ \frac{R1}{R2} \qquad (5.5)$$

or as :

$$R2 = 0.5\ R2\ \frac{I_{ss}}{I_{se}} \qquad (5.6)$$

Hence, the base-emitter voltages of devices Q2-5 will be equal. The size of the diode-connected BJT in Figure 5.12 is proportional to the ratio between the two currents I_{ss} and I_{se} to further ensure the equality of the base voltages of devices Q2-5. When one of the bit-line voltages drops, the current in the BJT connected to that line (and hence the current in that branch of the ECL amplifier) starts decreasing from its initial value of $0.5I_{ss}$. Meanwhile, the current through the other branch starts increasing. This imbalance in the currents of the two branches is converted to a small differential voltage at the sense-amplifier outputs. Due to the back-to-back connections between the two outputs through Q3 and Q4, a regenerative effect takes place which causes the total tail current to be quickly switched to the branch connected to the bit-line with the higher voltage. At the end of transition, the entire tail current I_{ss} will be flowing in one of the two devices Q3 or Q4. The sensing output voltages would remain constant even if the bit-line voltages changed (the latched-in state) until the sense-amplifier is reset by the SA signal becoming low. The voltage difference between the two sensing output would be :

$$\Delta Vo = I_{ss} \cdot R1 \qquad (5.7)$$

with the high output voltage being $VDD - V_{BE}$. This is another advantage of this sensing scheme; the outputs are already level-shifted by one V_{BE}. This gives more freedom to the designer to select the best topology for the next sensing stage.

Low-Power Embedded BiCMOS/ECL SRAMs 117

Figure 5.13 shows a complete column-sensing operation. This figure shows the Bs, Ws, WL, bit-line, and sensing output signals. The latched sense-amplifier correctly read and latched-in the data on the bit-lines which were only activated for 0.6 ns. The amplifier took less than 100 pS to latch-in the data on the bit-lines (with 75 mV bit-line's swing). When operated at its maximum speed (optimum Iss), this amplifier only consumes 11mW. At this optimum speed, the total delay from the WL to the sensing outputs was less than 400 pS for a load of 300 fF at the sensing outputs.

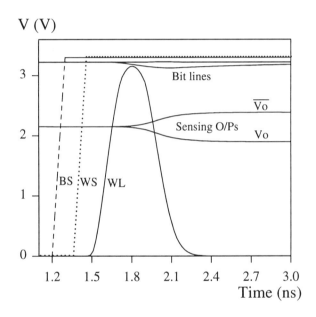

Figure 5.13 The simulated performance of the new column sensing technique.

Also, the way this sense amplifier is designed makes it very resilient to process and supply variations. This is due to the symmetrical nature of the circuit, the excellent matching of the bipolar devices, and the fact that R1 and R2, and the three NMOS current sources are multiples of one another. Assuming that all resistors are physically placed adjacent to each other, any variation in the value of R1 or R2 due to process will be consistent (the same percentage of change) for both resistors. The same applies to the three NMOS current sources. Hence equations (5.1) through (5.6) will always hold valid with process or supply voltage variations having minimum effect on the correct operation of the sense-amplifier.

5.5.1 Immunity to Bit-line Glitches

One concern with this sensing scheme is the possibility of sensing, amplifying, and latching-in false bit-line data (i.e. glitches). The regenerative nature of this sense-amplifier makes it prone to such erroneous behavior. To prevent this from happening, the regenerative effect can be slightly delayed, such that any bit-line glitch that has a duration less than a certain value is not latched-in. This can be achieved by adjusting R1 and R2, such that the initial value of the output voltage (given by equation (5.1) above) is kept slightly below the initial value of V1 and V2 given by equation (5.2). So the following condition will prevail :

$$Vo(t=0^+) \ = \ \overline{Vo}(t=0^+) \ = \ V1(t=0^+) \ - \ \Delta V_r \ = \ V2(t=0^+) \ - \ \Delta V_r \tag{5.8}$$

Hence the output regeneration will not start until one of the output voltages rises by at least ΔV_r. The larger ΔV_r is, the more resistant will be the sense-amplifier to glitches. However, delaying the output regeneration reduces the speed of the sense-amplifier and thus ΔV_r should be kept to a minimum value.

Figure 5.14 shows the effect of bit-line's glitches on the sense-amplifier outputs. Two bit-line glitches of a 50 mV amplitude with 100 ps and 200 pS durations, respectively were applied to the sense-amplifier inputs. For these results, the amplifier was designed such that ΔV_r value was about 100 mV. As Figure 5.14 shows, the temporary disturbances at the outputs were not latched-in. Although the gain of the amplifier was around 20, the output signals did not reach an amplitude of more than 80 mV before they quickly faded away. The response of the same amplifier to a correct bit-line signal of 75 mV amplitude and 500 pS has also been shown in Figure 5.14. The amplifier correctly read the bit-line signal and latched it in.

Low-Power Embedded BiCMOS/ECL SRAMs

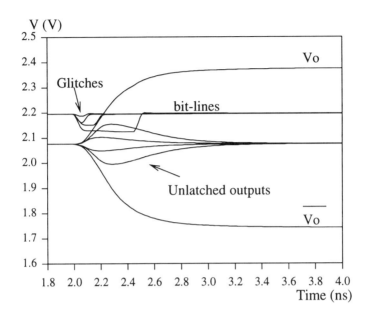

Figure 5.14 The response of the latched sense-amplifier to two 50 mV bit-line glitches of 100 pS and 200 pS durations. The response to a 75 mV and 500 pS bit-line read-out signal is also shown.

5.5.2 Performance Comparisons

A conventional high-speed ECL sense-amplifier with cross-coupled PMOS loads is shown in Figure 5.15. The cross-coupled load provides a dynamic load to the amplifier that enhances its speed. The equalization MOSFET limits the output signal swing and enhances the speed. The performance of this conventional sense-amplifier was compared to that of the new one. Figure 5.16 shows the delay versus power for both amplifiers. The new amplifier consistently achieved lower delay for the same power. The optimum delay of the latched sense-amplifier was 160 pS, 25% faster than the conventional one with the delay of 222 pS. At 8 mW, the new amplifier delay (250 pS) is still 20% lower than that of the conventional one. Again, although the delay savings may not seem very significant compared to the total access time of an SRAM, the power savings resulting from the new column activation/sensing technique are significant.

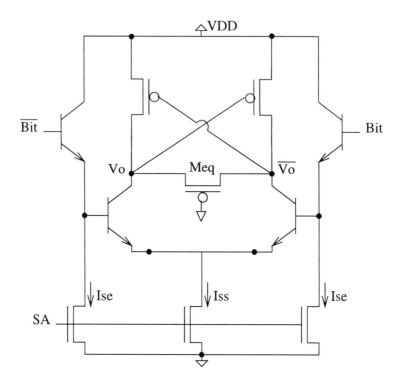

Figure 5.15 A conventional ECL sense-amplifier with cross-coupled PMOS loads.

The area of the new amplifier is slightly less than the conventional one. This is because the PMOS cross-coupled loads and equalization transistors have to be sized according to the value of the current I_{ss}. So, at a certain value of I_{ss}, the area of the conventional amplifier exceeds that of the new one.

Hence, the new amplifier correctly reads and latches-in the bit-line data requiring less time and area than the conventional sense-amplifier.

5.6 CHAPTER SUMMARY

In this chapter, the different front-end options of next generations of BiCMOS ECL SRAMs were evaluated in terms of speed and power. A 16 Mb[+] SRAM

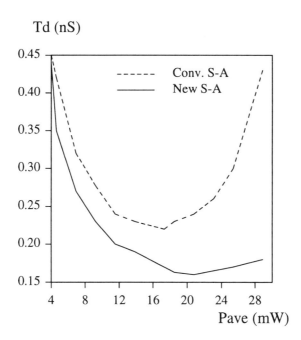

Figure 5.16 The delay of the two sense-amplifiers versus their average power.

front-end organization that is optimum in terms of speed was developed and verified. Several new circuit techniques that reduce the power consumption of different BiCMOS/ECL SRAMs sub-blocks while preserving or enhancing their speed were developed. These techniques include; low-power W-ORing and level-translation, self-resetting word-line decoding and driving, and unclocked bit-lines data sensing and latching. The performance of these new circuits was evaluated and compared to conventional circuits with similar functionality. It was shown that the new circuits do not have an area penalty and that they enhance the speed slightly. The estimated total power savings resulting from using the new circuit techniques presented in this chapter are between 22% to 43%.

REFERENCES

[1] A. R. Alvarez, "BiCMOS - Has The Promise Been Fulfilled?," In *BiCMOS Integrated Circuit Design: with Analog, Digital, and Smart Power Applications,* Ed. M. I. Elmasry, IEEE Press, 1994, pp. 15-18.

[2] A. G. Eldin, "An Overview of BiCMOS State-of-the-Art Static and Dynamic Memory Applications," In *BiCMOS Integrated Circuit Design: with Analog, Digital, and Smart Power Applications,* Ed. M. I. Elmasry, IEEE Press, 1994, pp. 257-267.

[3] S. Flannagan, "Future Technology Trends for Static RAMs," In *Digital MOS Integrated Circuits II: with applications to Processors and Memory Design,* Ed. M. I. Elmasry, IEEE Press, 1992, pp. 319-322.

[4] M. Takada, et al., "A 5-ns 1-Mb ECL BiCMOS SRAM," *IEEE J. of Solid-State Cir.,* vol. 25, pp. 1057-1062, 1990.

[5] C. T. Chuang, and D. D. Tang, "High-Speed Low-Power AC-Coupled Complementary Push-Pull ECL Circuit," *IEEE J. Solid-State Circuits,* Vol. 27, pp. 660-663, 1992.

[6] C. T. Chuang, et al., "High-Speed Low-Power ECL Circuit With AC-Coupled Self-Biased Dynamic Current Source and Active-Pull-Down Emitter-Follower Stage," *IEEE J. Solid-State Circuits,* Vol. 27, pp. 1207-1210, 1992.

[7] W. Wilhelm, and P. Weger, "2V Low-Power Bipolar Logic," *ISSCC Tech. Dig.,* pp. 94-95, 1993.

[8] H. Shin, "A Self-Biased Feedback-Controlled Pull-Down Emitter Follower for High-Speed Low-Power Bipolar Logic Circuits," *IEEE J. Solid-State Circuits,* Vol. 29, pp. 523-528, 1994.

[9] M. S. Elrabaa, M. I. Elmasry, and D. S. Malhi, "Low-Power BiCMOS Circuits For High-Speed Inter-Chip Communication," *IEEE J. Solid-State Circuits,* Vol. 32, April 1997.

[10] M. Yoshimoto, et al., "A 64Kb CMOS RAM with Divided Word Line Structure," *ISSCC Tech. Dig.*, pp. 250-251, 1989.

[11] T. Hirose, et al., "A 20nS 4Mb CMOS RAM with Hiirarchical Word Decoding Architecture," *ISSCC Tech. Dig.*, pp. 132-133, 1990.

[12] T. Chappell, et al., "A 2-nS Cycle, 3.8-ns Access 512-kb CMOS ECL SRAM with a Fully Pipelined Architecture," *IEEE J. Solid-State Circuits*, Vol. 26, pp. 1577-1585, 1991.

[13] M. S. Elrabaa, M. S. Obrecht, and M. I. Elmasry, "Novel Low-Voltage Low-Power Full-Swing BiCMOS Circuits," *IEEE J. Solid-State Circuits*, Vol. 29, pp. 86-94, 1994.

6
BICMOS ON-CHIP DRIVERS

6.1 INTRODUCTION

Several BiCMOS drivers have been developed throughout the years. Figure 6.1 shows some examples of these circuits. They can be divided into two categories: partial swing BiCMOS drivers (Figure 6.1(a) & (d)), and full-swing BiCMOS drivers (Figure 6.1(b), (c), (e), and (f)). When both pull-up and pull-down sections utilize bipolar devices the driver is usually classified as "BiCMOS" (Figure 6.1(a), (b), and (c)). When only the pull-up section includes a bipolar device and a simple NMOS device is used in the pull-down section the driver is classified as "BiNMOS" (Figure 6.1(d), (e), and (f)).

As VLSI technology scales down, the supply voltage also scales down. Although Bipolar devices do not usually suffer any significant loss in performance with scaling, BiCMOS circuits utilizing them do. They suffer losses in speed and output voltage swing. Meanwhile, the state-of-the-art high performance digital applications demand both speed and low power consumption at low supply voltages [1]. Also, at low supply voltages, fast and full-swing output waveforms become essential for both speed and static power dissipation of the driven CMOS gates. The speed of CMOS is greatly affected by the input slew rate [2] and if the input is of partial-swing, subthreshold leakage currents cause a non-zero static power dissipation.

Conventionally, there are two major circuit techniques used in BiCMOS full-swing circuits. The first technique utilizes shunting of the output BJT drivers in one of two configurations; base-emitter shunting (using MOS or resistors) (Figure 6.1(b) & (c)), or collector-emitter shunting (using MOS) [3]-[5].

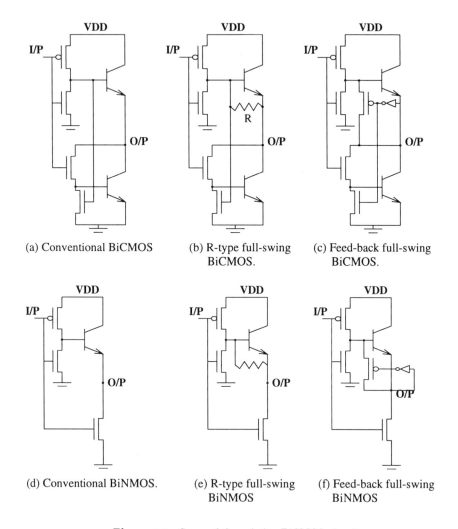

Figure 6.1 Some of the existing BiCMOS circuits.

The second technique utilizes PNP BJT's in one of two configurations. The first configuration is the emitter-follower configuration where a PNP is used in the pull-down section and an NPN is used in the pull-up section; both emitters connected to the output node [6]. In the second configuration, the common-emitter configuration, the PNP is in the pull-up section and the NPN is in the pull-down section [6]- [8]. This class of circuits requires a complementary BiCMOS process. Also to achieve full-swing at low supply voltages, the PNP, used in the common-emitter configuration, has to be saturated during transients [6, 8]. This results in an excess power consumption which is not used to charge/discharge the output (the emitter provides a large, unnecessary, base current).

A novel low-voltage low-power non-complementary full-swing BiCMOS circuit technique with superior performance over CMOS down to 1.5V was developed [9]. This new type of full-swing BiCMOS circuits is presented over the next sections.

6.2 THE NOVEL FULL-SWING BICMOS CIRCUIT TECHNIQUE

The new technique utilizes a conventional emitter-follower configuration combined with a positive capacitively-coupled feedback technique to achieve output swings very close to the rail supply for voltages down to 1.5 V. It has a more efficient usage of power and do not require PNP's or any other special processing.

The concept of operation of the novel circuits is presented next and compared to that of conventional BiCMOS circuits. This is followed by performance verification using both circuit and device simulations. Also, different building blocks that are used in high performance digital sub-systems were implemented using the novel circuit technique. The results of the evaluation of their performance as compared to that of optimized CMOS blocks with similar functionality are also presented. This performance comparisons includes delay, area, and power comparisons for several technologies and supply voltages. The design consideration of the new circuitry will also be presented.

6.2.1 Concept of Operation

6.2.1.1 Conventional BiCMOS Circuits

Conventionally, BiCMOS circuits utilized an emitter-follower configuration for the pull-up section as in Figure 6.2(a). In such circuits, the maximum output voltage (V_{omax}) is limited by the V_{BE} drop across the base emitter junction of the BJT driver. The HSPICE [10] transient circuit simulation results for the base and emitter voltages of such circuits, during output pull-up, are shown in Figure 6.2(b). As the input falls, the PMOS starts conducting, charging the base-emitter junction to $V_{BE(on)}$ and the Bipolar starts to conduct, raising the output node voltage. As the output rises, the base voltage also rises and approaches V_{DD}. The PMOS starts to turn off and the base current decreases and finally changes direction as shown in Figure 6.2(c). At this point, there is still a collector current due to the remaining base (minority carriers) charge. As a result, the output voltage continues to rise until it reaches a value that is slightly above $V_{DD} - V_{BE}$ by the end of transition. The voltage of the almost floating base increases above V_{DD} following the output voltage. However, the PMOS starts to conduct in the reverse direction, removing some of the extra base charge and bringing the base voltage back to V_{DD} (Figure 6.2(b)). The base charge, and hence the final value of the output voltage, depends on the value of the load capacitance C_L, the technology parameters, and the supply voltage [11]. For low C_L and/or scaled down technologies and supply voltages, V_{omax} is relatively lower.

6.2.1.2 The Novel Pull-Up Circuit

Referring to Figure 6.2(a), if the PMOS is turned off before the end of output transition, the charge leakage from the base would be significantly reduced, hence V_{omax} would increase. In fact, if the PMOS is turned off, about half of its channel charge would be injected into the base. This will boost the collector current for an additional period of time and increase V_{omax}. Also, if an additional source of charge is used to inject extra charges into the base, V_{omax} will increase even further.

Different pull-up circuits that employ the newly proposed technique are shown in Figure 6.3 combined with a simple NMOS pull-down circuitry. They use the above mentioned techniques to achieve full and fast output swing for supply voltages down to 1.5 V. In Figure 6.3, a feedback switch consisting of MP2 and MN1 is used to turn the PMOS (MP1) off by the end of output transition

BiCMOS On-Chip Drivers

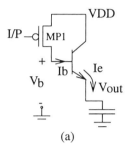

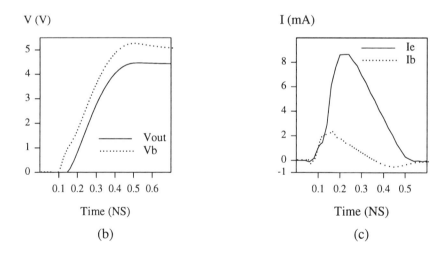

Figure 6.2 The conventional BiCMOS pull-up circuit; (a) schematics, (b) the transient response of the base and output voltages, and (c) the transient response of the base and emitter currents.

and a positive feedback capacitance C_{fb} is used as an additional source of base charge. The feedback switch, the feedback inverters in1 and in2, and C_{fb} will be referred to as the feedback circuit.

The novel circuits work as follows(referring to Figure 6.3): As the output voltage rises, a charge is stored in C_{fb}. When the output approaches the maximum value limited by the V_{BE} drop, the feedback circuit injects this charge back into the BJT base. The base voltage will rise, causing the BJT to continue to conduct which in turn makes the output voltage reaches the rail voltage. As C_{fb} starts injecting charges into the base, the feedback switch starts to turn the PMOS off hence minimizing charge losses. Turning the PMOS off at this point also leads to the injection of more charges into the BJT base. These extra charges come from the PMOS's channel charge which has to be dispersed when the PMOS is turned off. Since the source and drain voltages of the PMOS, at this point, are approximately equal, its channel charge will be divided equally between the source and drain nodes. Also, turning the PMOS off reduces the pull-down time since MN will have to discharge only the BJT base. Feedback timing is controlled by the sizing of the feedback inverters and will be discussed later.

The circuits in Figure 6.3(b)&(c) are non-inverting, while 2.3(a) is inverting. Circuit 2.3(a) however, would draw a substantial current from the driving gate (like MOS pass logic) during switching that may cause a dip in the input voltage. Circuit 2.3(b) is obtained from adding a CMOS gate to the input of 2.3(a), the simplest of all, while in 2.3(c), the CMOS gate drives the pull-down section, and only an NMOS block is used to drive the pull-up section to reduce the parasitics at the source node of MN1. A merged BiPMOS device was used for both the NPN and MP1 for these two reasons; 1) To avoid the loss of additional charge in the MP1 drain-substrate diode when the base node voltage goes above V_{DD}, 2) To reduce the overall area. The dotted section of Figure 6.3(a) is shown in Figure 6.4. The merged device used had a width of $20\mu m$ and the depths of emitter, base, buried layer, and P-substrate were $0.01\mu m$, $0.11\mu m$, $0.6\mu m$, and $1.0\mu m$, respectively. The drain depth was about $0.08\mu m$ and the effective metallurgical channel length was $0.38\mu m$. The different doping profiles were adjusted to give typical device parameters of a $0.5\mu m$ BiCMOS technology. Two one-dimensional, two-dimensional, and quasi-two-dimensional DC device simulations were used in this process. The Bipolar gain was enhanced by reducing the base current by decreasing the surface recombination velocity at the emitter contact.

BiCMOS On-Chip Drivers

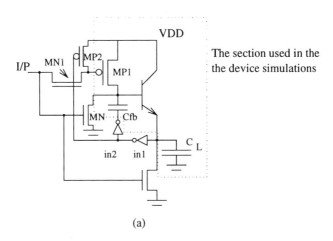

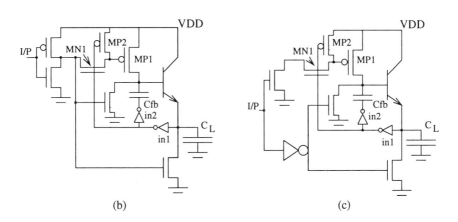

Figure 6.3 Different implementations of the novel BiCMOS pull-up circuits utilizing the positive dynamic feedback.

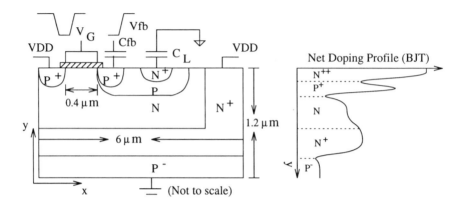

Figure 6.4 A cross-section of the merged BiPMOS device used in the transient device simulations.

6.2.2 The Verification of Operation

The operation of the novel pull-up circuits was verified using both circuit and device simulations.

6.2.2.1 Circuit Simulations

In order to simulate the merged structure using HSPICE , the PMOS MP1 drain diode area (connected to the BJT base) was set to zero. This accurately models the merging of the base and drain layers, however this does not take care of all of the effects of merging on the PMOS N-well currents. This is why further device simulations of the merged structure were performed, as will be demonstrated below.

The circuit simulations showed that the novel circuits can achieve swings very close to the rail voltage for different technologies and supplies, as shown in Figure 6.5. The HSPICE parameters of those generic BiCMOS technologies are in Table 6.1. The output is fast for the whole duration of transients, i.e., there is no slow portion at the end of transition as in circuits that utilize MOS shunting (e.g. in [5]). Also, Figure 6.6 shows the emitter current of the BJT, during the pull-up and pull-down transients, at a frequency of 250 MHZ. During pull down, the emitter current is zero, indicating that the excess minority carriers trapped in the collector, due to saturation, do not affect the

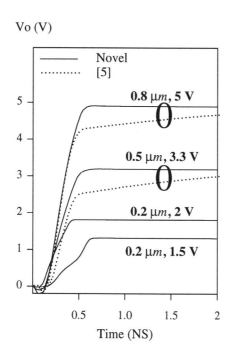

Figure 6.5 The pull-up transient response of the novel circuit for, several technologies and supply voltages (solid lines), and for $C_L = 0.3pF$. For the (0.8 μm, 5V) and (0.5 μm, 3.3V) technologies, the dotted lines represent the response of the circuit in [5] with an MOS output shunt.

pull-down transients. This is because unlike the circuit in [8], where a PNP collector is driving the output, an NPN emitter is driving the output.

6.2.2.2 *Transient Device Simulations*

A two-dimensional transient device simulator TRASIM [12], that was developed from DC device simulators for MOS and Bipolar devices [13, 14], was used to check V_{omax} and the effects of turning off the PMOS and the feedback capacitance on it. The structure in Figure 6.4 was simulated. The gate voltage, V_G, was changed from V_{DD} to zero and back to V_{DD} to simulate the feedback switch. As V_G increases back to V_{DD}, V_{fb} is increased from zero to V_{DD}. The start of the feedback action is defined as the time when both V_G and

V_{fb} change from low to high. The initial values of the base and emitter voltages were set to zero. Figure 6.7 shows that turning the PMOS off, increases the swing significantly, and adding the feedback capacitance increases it even further. Although this figure might undermine the importance of the feedback capacitance, since it only increased the swing by about 0.2 V, it should be noted that the subthreshold slope of the state-of-the-art MOSFET's is about 90 mV/decade. This means that the use of C_{fb} will reduce the leakage currents in CMOS gates driven by the output by about two orders of magnitude. Also Figure 6.7 shows that the output voltage does not have a 'slower' portion, instead its slope actually increases after the start of feedback.

Device simulations were also used to check for latch-up resulting from a parasitic PNP in the merged BiPMOS devices that was reported to be a potential source of latch-up [15]. There are two parasitic PNP's to be considered. One is under the PMOS gate between the base and source, and the other is under the base between the base and the substrate, as shown in Figure 6.8.

From the two-dimensional (2-D) hole distribution at different points of time during transients, the following could be noticed :-

CMOS				Bipolar (min. size)				
$L_{eff}(\mu m)$	$V_{th}(V)$ NMOS PMOS	$t_{ox}(nm)$	C_j $(fF/\mu m)$	β	$\tau_F(PS)$	I_K (mA)	$C_c fF$	$C_s fF$
0.8	0.8 -0.8	15	0.4	100	10	2.5	9.5	25
0.5	0.56 -0.6	12	0.6	100	7	1.8	7.5	18
0.2	0.35 -0.35	7	1	100	4	1.0	3.5	8.5

Table 6.1 HSPICE parameters of the three generic BiCMOS technologies.

BiCMOS On-Chip Drivers

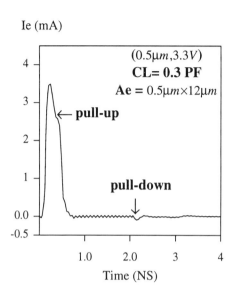

Figure 6.6 The emitter current of the pull-up BJT during pull-up and pull-down at a 250 MHz frequency.

1. There is a PNP between the source and the base. It enhances the performance before the feedback by increasing I_b. However, after the start of feedback, it reverses direction, saturates (Figure 6.9(b)) and removes some of the base charge, hence, decreasing the voltage swing. This PNP could be eliminated using special layout techniques for the merged structure [15].

2. Also, after the start of feedback, holes are injected into the collector, due to saturation. However, hole injection into the substrate remains insignificant, as Figure 6.9(b) shows.

3. By the end of transition, the hole concentration in the N-well, under the gate and base, falls by a few orders of magnitude (Figure 6.9(c)). This means that there are no longer any parasitic PNP's and hence no runaway conditions.

several orders of magnitude, as in Figure 6.10(b). This The pull-down circuit will not be affected by the extra base charge (excess electrons in the base), confirming the results obtained from circuit simulations.

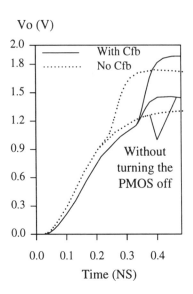

Figure 6.7 The transient response of the novel pull-up circuit with and without turning the PMOS off. Also the effect of C_{fb} is shown.

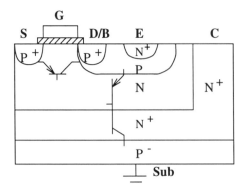

Figure 6.8 The two parasitic PNP BJTs in the merged BiPMOS structure.

6.3 PERFORMANCE COMPARISONS

In this section the performance of three types of circuits implemented using the novel circuit techniques will be evaluated for several design conditions. Their performance in terms of speed and power would be compared to other circuits with similar functionality. The implemented circuits are; *simple buffers, AND gates, and master-slave D latches.*

6.3.1 Simple Buffers

6.3.1.1 Delay Comparison

The delay time T_D, measured from the instant the input $= V_{DD}/2$ to the instant the output $= V_{DD}/2$, was calculated from results of HSPICE simulations of the circuits in Figure 6.3 as a function of the load capacitance C_L. The $(0.2\mu m, 2\ V)$ technology parameters were used in these simulations. Also, T_D of an optimized CMOS buffer with the same input capacitance as the three novel BiCMOS buffers was evaluated for the same technology. The CMOS buffer was limited to one or two stages only (depending on the value of C_L), for practical area considerations. Also, for all circuits, the devices were sized such that the rise and fall times of the circuit were approximately equal. Figure 6.11 shows the result of this comparison.

The novel circuits outperformed CMOS down to below 0.2 PF (Fanout of less than 4) load capacitance, something that was not achieved by any other reported BiCMOS circuit at 2 V. The novel circuits start to lose their full-swing capabilities at loads below $0.2pF$ due to the very fast output rise time, which in turn cause the feedback circuitry to turn-off the PMOS (and hence the BJT) prematurely. This means that these novel buffers have to be redesign for low-loads applications (slower feedback).

As Figure 6.11 shows, the speed difference between the novel circuits and the CMOS buffer start increases as C_L increases, but it then starts to decrease and finally the optimized CMOS buffer becomes faster. However, at such a point the optimized CMOS buffer chain becomes excessively larger. Similar behavior was recently reported for conventional BiCMOS buffers when compared with optimized CMOS buffer chains [16, 17].

Two pull-down circuits were tested in conjunction with the novel pull-up circuit, in addition to the simple NMOS pull-down circuitry. One circuit is a novel one

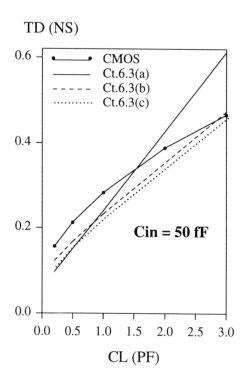

Figure 6.9 The average delay of the circuits in Fig. 6.3 compared to the optimized CMOS buffers and using the $(0.2\mu m, 2V)$ technology HSPICE parameters.

and is similar to the novel pull-up circuit, but does not have the feedback capacitance, and the other is similar to the one in [8], as shown in Figure 6.12. The delays of these two circuits, that of circuit 2.3(a) (with the NMOS pull-down), and the optimized CMOS buffer, are shown in Figure 6.13. The novel pull-up/pull-down circuit (ct. 2.12(a)) achieved the highest speed-up over CMOS for higher load capacitance. The circuit with NMOS pull-down (ct. 2.3(a)) achieved good speed-up at lower load capacitance and had the least area among the three BiCMOS circuits.

BiCMOS On-Chip Drivers

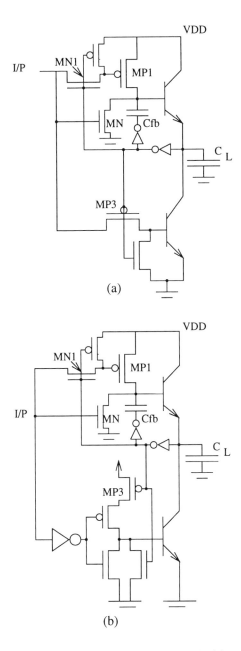

Figure 6.10 The novel pull-up circuit combined with; (a) a novel pull-down circuit, and (b) a pull-down circuit similar to the one in [6].

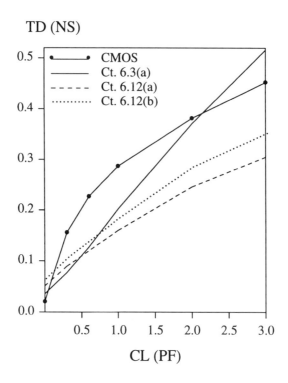

Figure 6.11 The average delay of the circuits in Fig.6.12 and Fig.6.3(a) compared to CMOS for the $(0.2\mu m, 2V)$ technology

6.3.1.2 Power Comparison

The average power dissipation of the three BiCMOS circuits, 2.3(a), 2.12(a), and 2.12(b) and that of the CMOS buffer at 100 MHZ is shown in Figure 6.14 as a function of C_L. For these results the CMOS buffer consisted of only a single stage. Circuit 2.3(a) has the least power dissipation. The two other BiCMOS circuits have power consumptions that are very close to that of the CMOS buffer, especially circuit 2.12(a).

BiCMOS On-Chip Drivers

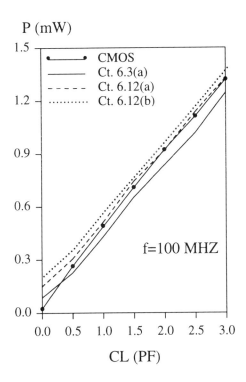

Figure 6.12 Power vs C_L for the circuits of Figures 6.12 and 6.3(a) compared to CMOS for the $(0.2\mu m, 2V)$ technology.

6.3.1.3 Delay versus Supply Voltage

The delay of circuit 2.12(a) was evaluated as a function of the supply voltage, V_{DD}, and compared to that of the two stage CMOS buffer for the three BiCMOS technologies, as shown in Figure 6.15. For the sake of a fair comparison, both the area and the input capacitance of the two circuits were kept approximately equal. The value of C_L used was approximately equivalent to a fanout of 4. For the $0.8\mu m$ technology the novel BiCMOS circuit did not operate below 2 V supply voltage. However, it outperformed the CMOS buffer down to that voltage. As for the $0.5\mu m$ technology, the novel circuit outperformed CMOS down to about 1.7 V. At $0.2\mu m$ it even outperformed CMOS down to 1.5 V. These results were not achieved by any previously reported BiCMOS circuits that do not utilize the highly expensive PNP's.

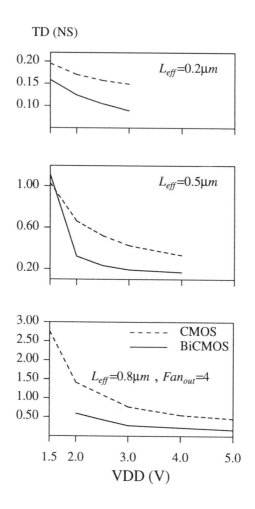

Figure 6.13 Delay vs supply voltage of circuits 6.12(a) and the CMOS buffer for the three BiCMOS technologies.

6.3.2 AND Gates

A multi-input AND gate was implemented using the novel circuit technique with slight modifications to increase the speed, Figure 6.16. As this figure shows, the feedback switch is slightly different. MP2 is connected as a load and the NMOS MN1 was placed at the bottom of the NMOS logic block so that a PMOS logic block, in the pull-up section, becomes unnecessary and hence was discarded. There is no static power dissipation since the feedback turns the MN1 off before the end of transition. This configuration saves area and enhances speed by reducing the parasitic capacitance at the gate of the BiPMOS merged device. The feedback capacitance was dropped out since this circuit was intended for the $0.5\mu m$ and 3.3 V technology. Turning the PMOS off was sufficient to make the output voltage reach 3.1 V. This value of V_{omax} did not decrease as the number of inputs was increased up to 7. Also, an NMOS base-emitter shunt is used to discharge the NPN base. Since V_{th} of the NMOS is smaller than $V_{BE(on)}$, the BJT will be turned off at the end of pull-up and would remain off during the pull-down. The pull-down section is similar to that in [8] with a few modifications. The NMOS MN2 is now diode connected and the unnecessary NMOS logic block was removed to reduce area and parasitic capacitance at the base of the BJT.

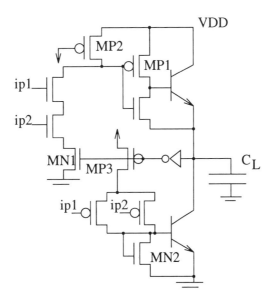

Figure 6.14 An AND gate implemented using the novel circuit technique.

The pull-down section of the AND gate operates as follows; starting with high inputs and output, as one or more input changes to low, the corresponding PMOS in the pull-down section turns on. Hence turning the Bipolar on which starts discharging the output node. This continues until the feedback inverter turns MP3 off and MN2 starts discharging the Bipolar base and turning it off. Meanwhile, the shunting NMOS in the pull-up section keeps the BJT off. At the end of the transition, the feedback switch in the pull-up circuit is turned on and the circuit is ready for a pull-up transition. The operation of the pull-up section is similar to that of the simple buffer explained in section II. However, in this circuit, the NMOS shunting the BJT is turned on at the end of transition such that the circuit is ready for the next pull-down transition.

The sizing of transistors MP2 and MN2 is very important. MP2 should be small enough not to slow the NMOS logic chain, yet large enough to prevent or reduce glitches at the BiPMOS gate node. MN2 should be small enough not to slow the parallel PMOS logic block from turning the BJT on, yet large enough to discharge the BJT in adequate time (depending on the frequency).

The speed-up and the area ratio between the novel BiCMOS AND gate and a CMOS NAND with equal input capacitance for the $(0.5 \mu m, 3.3V)$ technology are shown as a function of the number of inputs in Figure 6.17. The speed-up is defined as :

$$Speed - Up\ Factor\ =\ \frac{Delay\ of\ CMOS\ Gate}{Delay\ of\ Novel\ BiCMOS\ Gate} \quad (6.1)$$

This figure shows that for a single input, the speed-up factor is about 1.7 and the area ratio is about 4.3. For the six input gates, the speed-up factor reaches a maximum of 2.7 and the area ratio drops to about 1.6. Hence in macrocells (e.g. adders, ALU's ..etc.) implemented using the novel techniques, as the number of inputs per logic gate increases, the overall speed-up over the CMOS implementation increases while the relative area penalty decreases. This is confirmed by results reported in [1] where the speed-up of a BiCMOS carry lookahead adder over a CMOS one increased as the average fan-in per logic gate increased. The technique reported here, however, in addition to being faster than CMOS, would consume less area than those in [1], rendering it more attractive.

6.3.3 Master-Slave D Flip-Flops

A master-Slave D flip-flop that utilizes a single-phase clocking scheme implemented using the novel circuit techniques is shown in Figure 6.18. The single-phase clock operation is very essential for future RISC processors applications [1]. A clocked inverter is used to hold the output when the inputs are

BiCMOS On-Chip Drivers

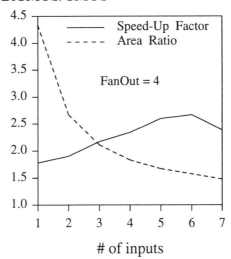

Figure 6.15 The speed-up and area ratio between the novel BiCMOS AND and the CMOS NAND for the $(0.5\mu m, 3.3V)$ technology and as a function of the Fan_{in}.

disabled by either the clock (CK) or the inverted clock $(\overline{CK})$ [1]. The pull-up section is similar to that of the AND gate, except for the PMOS MP3 at the input used to prevent glitches at the gate node of the BiPMOS device. These glitches may occur when the input is low and the clock goes high. This also eliminates the need for a clocked inverter at the input. Hence the area and the delay are reduced by eliminating the unneeded series PMOSFETs. The PMOS in the feedback switch is now controlled by the feedback inverter. The NMOS shunting the BJT is controlled by by the pull-down section. Unlike the AND gate or the gate in [1], a new arrangement is used for the pull-down section. It consists of a regular feedback switch connected in series with another switch that is controlled by the input and the clock. This arrangement eliminates the need for staking three PMOSFET's on top of each other, which would slow the circuit and hinder its low voltage operation.

This flip-flop, like the one in [1], does not suffer in performance if there is a clock skew between CK and $\overline{CK}$. However, it will have smaller area and faster output response even at low supply voltages.

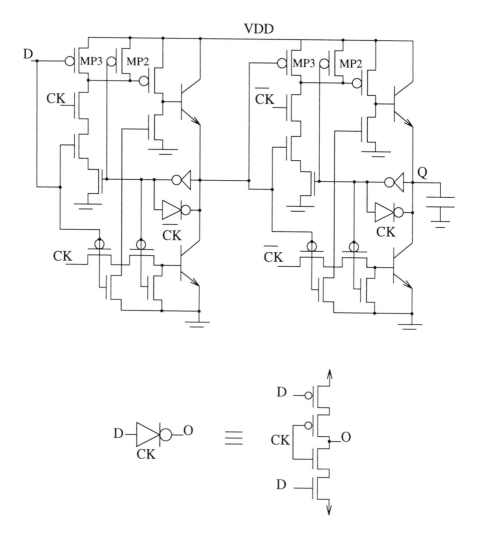

Figure 6.16 A Master-Slave latch implemented using the novel circuit technique.

The performance of the novel BiCMOS D flip-flop was compared to that of a CMOS single-phase clocked D flip-flop with the same input capacitance and an approximately equal area for several supply voltages. Using the $0.5\mu m$ BiCMOS technology and a Fan_{out} of 4, the write and total delays of both circuits are reported in Figure 6.19 as a function of the supply voltage. The write time, is the time required to transfer the data from the input to the output of the master latch. The total delay time is the time required to transfer the data from the input to the output of the slave latch. The novel flip-flop did not only outperformed the CMOS one by almost a factor of 2 in total delay down to 1.5 V supply, but it also had a smaller write time.

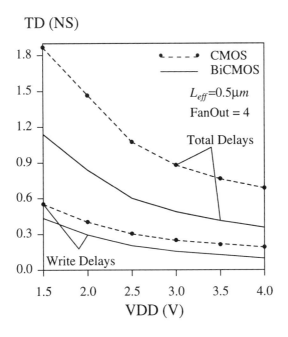

Figure 6.17 The write and total delays of the BiCMOS and CMOS latches vs the supply voltage.

Hence the new circuit techniques can be used to implement buffers, logic gates, and master-slave latches that exceed the speed of CMOS for equal input capacitance and silicon area, and for different technologies and supply voltages.

6.4 DESIGN OF THE FEEDBACK CIRCUITRY

The feedback circuit consists of three parts; 1) the feedback switch, MN1 and MP2 in Figure 6.3, the feedback capacitance C_{fb}, and the feedback inverters in1 and in2 (Figure 6.3). The size of MN1 equals that of the other NMOS transistors in the NMOS series logic block at the input, which are sized according to the input capacitance (loading on the driving gate) specifications. MP2 is minimally sized such that it does not turn MP1 off prematurely or too late. In either case this would lead to a smaller output swing. The sizing of MP2 is strongly coupled to that of the feedback inverter in1. So the NMOS in in1 and MP2 are sized simultaneously such that the feedback switch is turned off when the output is about $V_{DD} - V_{BE(on)}$. And if in1 is also used to control the pull-down section (as in the circuits in Figures 2.12, 2.16, and 2.18) its PMOS should be sized such that the pull-down feedback switch is also turned off at the proper time. The discharging NMOSFETs, used to discharge the bases of the pull-up/down BJT's, should be sized minimally and according to the frequency of operation such that they discharge the BJT's base in adequate time, yet do not load the circuit significantly. Inverter in2 is sized depending on the value of C_{fb}, the larger C_{fb} is, the larger the PMOS in in2 should be. However, the width of the NMOS in in2 is set close to the minimum width.

The value of C_{fb} depends on many factors, such as the loading, the technology, and the supply voltage. C_{fb} should be able to hold enough charge to charge the NPN base to about 0.7 V above V_{DD} and supply the base charge needed to make the NPN continue conducting till the output reaches V_{DD}. For a base-collector junction capacitance C_{bc}, the minimum C_{fb} required to charge the base to $V_{DD} + 0.7V$ would be :

$$C_{fb\ (min)} = \frac{0.7\ C_{bc}}{V_{DD}} \qquad (6.2)$$

Noting that C_{fb} is also assisted by the charge injected by the turning off of the PMOS MP1, C_{fb} does not need to be much greater than the value of $C_{fb\ (min)}$. This is important since increasing the value of C_{fb} decreases the output slew rate and hence increases the delay, as was shown in Figure 6.7. This means that for a supply voltage of 3 V and a C_{bc} of about 20 fF, a C_{fb} of 15 fF would probably be sufficient for the circuit to achieve a full swing without compromising the speed significantly. For very high loads, C_{fb} would have to be increased above the value of $C_{fb\ (min)}$, especially if the charging PMOS is not very large. The amount of charge supplied to the NPN base by the PMOS when it turns off, ΔQ, could be roughly estimated as half of the total channel charge under the gate, i. e. :

$$\Delta Q = \frac{1}{2} C_{ox} V_{DD} \qquad (6.3)$$

where C_{ox} is the gate oxide capacitance of MP1. If this capacitance is about 40 fF and V_{DD} is 3 V, ΔQ would supply the NPN with an average base current of about 0.6 mA for a hundred picosecond, a current that is usually sufficient to achieve full swing. It should be noted, however, that a portion of that current would be injected into the collector of the now saturated NPN. Hence the circuit designer should not rely solely on ΔQ to achieve full swing, especially at lower supply voltages as was shown in Figure 6.7.

The re-design of the feedback circuit as the technology scales is not straight forward. This is because supply voltage scaling does not usually follow that of the horizontal dimensions, hence the value of $C_{fb\ (min)}$ in equation (6.2) above will not remain constant with scaling. However, a simple analysis reveals the following; although both C_{bc} and V_{DD} scale down in different proportions that may cause the value of $C_{fb\ (min)}$ to increase, the amount of charge required to keep the NPN conducting will become smaller with scaling. This is because the base will be shallower, its area will be smaller, and the collector doping will be higher leading to smaller base charge, smaller leakage surface for the injected charge, and less parasitic PNP latch-up, respectively. This means that the required value of C_{fb} will probably not increase significantly with scaling. For this work, the values of C_{fb} used were around $20fF$, $30fF$, and $50fF$ for the $(0.8\mu m, 5V)$, the $(0.5\mu m, 3.3V)$, and the $(0.2\mu m, 2V)$ technologies, respectively.

6.5 CHAPTER SUMMARY

A novel full-swing BiCMOS circuit technique with superior performance over CMOS down to 1.5V was implemented using a conventional non-complementary BiCMOS process. The pull-up section is based on a capacitively-coupled feedback circuit. Both circuit and transient device simulations were used to verify and confirm the correct operation of the new circuits. Different pull-down options were examined and compared. Several circuits with various complexities were implemented using the novel circuit technique. These include simple buffers, logic gates, and master-slave latches. Their performance, regarding speed, area and power, was compared to that of CMOS for several different technologies and supply voltages. A design procedure for the feedback circuit and the effects of scaling on that procedure was summarized.

REFERENCES

[1] K. Yano, et al., "3.3-V BiCMOS Circuit Techniques for 250-MHz RISC Arithmetic Modules," *IEEE J. of Solid-State Cir.*, vol. 27, pp. 373-381, 1992.

[2] N. Hedenstierna and K. Jeppson, "CMOS Circuit Speed and Buffer Optimization," *IEEE Trans. on Computer-Aided Design,* Vol. CAD-6, pp. 270-281, 1987.

[3] S. H. Embabi, A. Bellaouar and M. I. Elmasry, "Digital BiCMOS Integrated Circuits Design," Kluwer Academic Publishers, 1993.

[4] Y. Nishio, et al., "A BiCMOS Logic Gate with Positive Feedback," *ISSCC Tech. Dig.*, pp. 116-117, 1989.

[5] H. Hara, et al., "0.5 μm 2M-Transistor BipnMOS Channelless Gate Array," *ISSCC Tech. Dig.*, pp. 148-149, 1991.

[6] H. Shin, "Performance Comparison of Driver Configurations and Full-Swing Techniques for BiCMOS Logic Circuits," *IEEE J. of Solid-State Cir.*, Vol. 25, pp. 863-865, 1990.

[7] S. H. Embabi, et al., "New Full-Voltage-Swing BiCMOS Buffers," *IEEE J. of Solid-State Cir.*, Vol. 25, pp. 150-153, 1991.

[8] M. Hiraki, et al., "A 1.5V Full-Swing BiCMOS Logic Circuit ," *ISSCC Tech. Dig.*, pp. 48-49, 1992.

[9] M. S. Elrabaa, M. S. Obrecht, and M. I. Elmasry, "Novel Low-Voltage Low-Power Full-Swing BiCMOS Circuits," *IEEE J. Solid-State Circuits,* Vol. 29, pp. 86-94, 1994.

[10] *HSPICE User's Manual,* Meta-Software, Inc., Campbell, CA, 1990.

[11] T. Arnborg, "Performance Predictions of Scaled BiCMOS Gates Using Physical Simulation," *IEEE J. of Solid-State Cir.*, Vol. 27, pp. 754-760, 1992.

[12] M. S. Obrecht, M. I. Elmasry, and E. L. Hesselle, "TRASIM: Compact and Efficient Two-Dimensional Transient Simulator for Arbitrary Planar Semiconductor Devices," *IEEE Trans. on CAD* vol.14, pp.447-458, 1995.

[13] M. S. Obrecht, "SIMOS - Two-Dimensional Steady-State Simulator for MOS Devices," *Solid-State Electronics,* Software Survey Section, vol. 32, No. 6, 1989.

[14] M. S. Obrecht, and J. M. Teven, "BISIM - A program for Steady-State Two-Dimensional Modeling of Various Bipolar Devices," *Solid-State Electronics,* Software Survey Section, v.34, No.7, 1991.

[15] H. Momose, *et al.,* "Characterization of Speed and Stability of BiNMOS Gates With A Bipolar and PMOSFET Merged Structure," *iEDM Tech. Dig.,* pp. 231-234, 1990.

[16] M. S. Elrabaa and M. I. Elmasry, "Design and Optimization of Buffer Chains and Logic Circuits in a BiCMOS Environment," *IEEE J. of Solid-State Cir.,* vol. 27, pp. 792-801, 1992.

[17] M. S. Elrabaa and M. I. Elmasry, "Optimization of Digital BiCMOS Circuits, An Overview," *Proc. of the 35th Midwest Sym. on CAS,* pp. 571-574, 1992.

7
INTER-CHIP LOW-VOLTAGE-SWING TRANSCEIVERS

7.1 INTRODUCTION

The BiCMOS technology has proven to be an excellent workhorse for telecommunication applications [1]. This is due to its excellent digital and analog capabilities as well as the variety of I/O's it offers. Consequently, this made the design process, using the BiCMOS technology, a very flexible one. However, lately due to the ever increasing complexity of telecommunication switches, the need has arisen for novel circuits with low on-chip power consumption. This is particularly important in two classes of circuits which conventionally consumed the most power; signal translation circuits, and I/O circuits. Moreover, since the DSP portions of telecommunication chips are usually implemented using CMOS logic, the supply voltages will have to be scaled down for future submicron and deep submicron BiCMOS technologies to maintain a high reliability for the CMOS circuits. Also, many of the recently reported low-voltage-swing driver circuits have a compatibility problem. These circuits range from reduced-swing CMOS [2], and CMOS pseudo-ECL or CMOS 100K ECL [3, 4], to CMOS GTL [5]. While the CMOS reduced-swing transceivers have limited speed, the CMOS true or pseudo ECL are complicated to design and have high power consumption, and the GTL requires different reference and termination voltages. Each of these transceivers as well as the true or pseudo Bipolar ECL or CML transceivers requires a different termination voltage and hence the incompatibility problem arises. This means that signal conversion parts as well as multiple termination and reference voltages would be required in systems using parts with different transceiver types, thus increasing the overall system cost and complexity.

The circuits described in this chapter represent attempts to solve the power and incompatibility problems simultaneously through three parallel venues :-

1. Power supply voltage reduction,
2. Power-Delay optimization,
3. Novel low-power circuit techniques.

In this chapter several novel low-power low-voltage-swing BiCMOS circuits that were designed to be used in chip-to-chip communications are presented. Two novel dynamic circuit techniques that enhance ECL/CML drivers speed, while significantly reducing their power consumption, are presented. The newly developed techniques are; the dynamic active-pull-down (DAPD) buffering/level shifting of the drivers input signals and the dynamically-controlled charge-pumped (DCCP) current sourcing. The effects of these techniques on ECL/CML drivers speed/power are also presented. Their performance is compared to that of conventional circuit techniques with similar functionality for a supply voltage of 3.3V.

A universal BiCMOS low-voltage-swing transceiver (driver/receiver) with low on-chip power consumption is also presented. Measurement results of test circuits fabricated using a $0.8\mu m$ BiCMOS technology are also included.

A Northern Telecom ($0.8\mu m$, $5V$) BiCMOS technology combined with a supply voltage of 3.3 V (the next standard voltage for digital ASICS) was used in the simulations, fabrication, and testing of the new circuits. The key device parameters of the technology are in Table 7.1.

7.2 LOW-POWER ECL/CML DYNAMIC CIRCUIT TECHNIQUES

Typically, ECL/CML circuits are differentially driven, and for high speed operation, the level-shifters/buffers of the input signals (usually an emitter-follower circuit) are biased at very high levels of DC power. Also, the power consumed by the current switch is constantly high no matter what is the state of the output. For single-ended drivers this is wasteful of power. Different dynamic circuit techniques that reduce the power consumption of ECL/CML logic circuits while maintaining or increasing their speed were recently reported [6]- [9]

β	I_K	τ_F	R_B	R_C	C_{eb}	C_{bc}	C_{cs}
97	5 mA	12 pS	330	87	15 fF	24 fF	45 fF

(a) The Bipolar (1x) parameters

	V_{th}	T_{ox} (nm)	I_{DSmax} (mA/U)	$C_{S/D}$ (fF/U^2)
NMOS	0.8 V	17.5	0.375 *	0.25
PMOS	-0.85 V	17.5	0.190 *	0.45

* For VGS= 5 V

(b) The MOS parameters

Table 7.1 NT's (0.8 μm, 5V) BiCMOS Technology parameters.

(Figure 7.1). However, these techniques were intended for Bipolar VLSI logic applications and are not suited for the high-speed high-drivability ECL/CML output drivers applications.

Two novel circuit techniques that solve the above problems are presented below. These techniques are the dynamic active-pull-down (DAPD) buffering/level shifting of drivers input signals and the dynamically-controlled charge-pumped (DCCP) current sourcing. The performance of the new circuit techniques is compared to the conventional circuits for a supply voltage of 3.3V.

For this work the ECL driver is assumed to have a 50Ω termination to a termination voltage $VT = VCC - 2V$ while the CML driver has a 50Ω termination to VCC as shown in Figure 7.2. For both drivers an output load capacitance of 2 pF and a fanout of 1 were used. The inputs are either driven by the outputs of a conventional differential ECL buffer directly, or the outputs of the ECL buffer were first level-shifted by one of the level shifting circuits of Figure 7.3. The current source in the current-switches of the drivers was implemented using one of the circuits in Figure 7.4.

156 CHAPTER 7

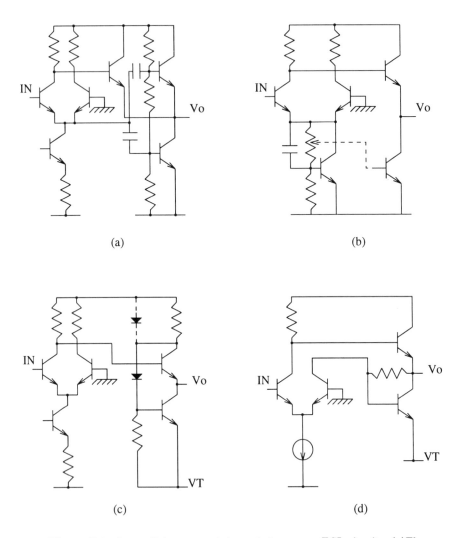

Figure 7.1 Some of the reported dynamic low-power ECL circuits; (a)The AC-PP-ECL of [6], (b) The AC-CS-APD-ECL of [7], (c) The FPD-ECL of [9], and (d) The active push-pull ECL of [8]

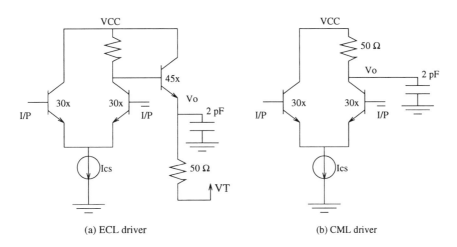

Figure 7.2 The two Bipolar current-mode drivers ; a) The ECL driver, and b) the CML driver.

7.2.1 The DAPD Level Shifter

The sizes of the BJT's at the inputs of ECL/CML drivers are much larger than those of the gates driving them. Hence, an emitter-follower stage (Fig. 3(a)) is usually used for level shifting and buffering of the input signals of the drivers. While this circuit performs well in terms of buffering it consumes a large amount of standby power especially if it is designed to operate at very high frequencies. This is due to the non-active pull-down operation of this circuit which is achieved via a resistor. Hence to balance the rise and fall delays at high frequencies the resistor value is reduced, thus increasing the DC power.

In the new circuit of Figure 7.3(b) the pull-down BJT, Q_2, is biased by the biasing network to have a low stand-by current. When the input $\overline{I/P}$ goes from 'Low' to 'High' (and I/P goes 'Low'), the capacitor C_D, injects a relatively large charge into the base of Q_2 boosting its current to quickly turn-off the input of the driver. Meanwhile, the biasing network acts as a feedback; as node 1 voltage rises slightly, increasing the current in Q_D, the collector current of Q_B increases quickly bringing the base voltage of Q_2 to its original value. When $\overline{I/P}$ goes 'Low', the base of Q_2 goes low instantaneously, reducing its collector current so that Q_1 can quickly turn on the input of the driver. The biasing network would then decrease Q_D emitter's current such that the collector current of Q_1 goes back to its original value.

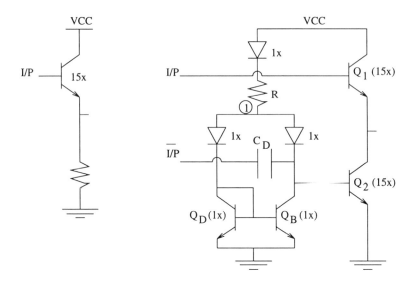

(a) A conventional level-shifter (b) The novel low-power level-shifter

Figure 7.3 The level shifting circuits; a) The conventional Emitter-Follower circuit, and b) The novel DAPD circuit.

The stand-by current through Q_1 and Q_2 is set by the current through Q_D and the area ratio between Q_2 and Q_D. The current through Q_D is $= \frac{V_{CC} - 3 \cdot V_{BE}}{R}$. The size of $Q_{1,2}$ is determined by the loading effect on the driving gate, and hence it was set to 15x. R then was adjusted accordingly to produce the required bias current. The bias current is adjusted for approximately equal output voltage rise and fall times. This was achieved by making the output transit charging and discharging currents equal. The value of the capacitance C_D is set to the maximum value that does not load the driving gate and is also subject to area constrains. It was found that increasing it beyond a certain value (about 0.5pF) does not significantly enhance the performance any more.

Figure 7.5 shows the collector current of Q_2 during the pull-down and pull-up transitions. It is shown there that Q_2 sinks a large current (> 10 mA) during the pull-down transition and a much smaller current (< 1 mA) during the pull-up transition. This leads to the extremely fast turn-on/off of the driven Bipolar transistors (in the output drivers). This figure also shows how the biasing network quickly (in less than 1 nS) brings the collector current of Q_2

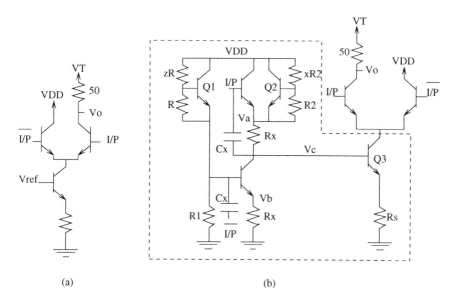

Figure 7.4 The two current source implementations; a) The conventional current source, and b) The novel DCCP current source.

to its moderately low steady-state value no matter what the final state of the output is.

The effects of the novel level shifting circuit on the performance of the two types of drivers (of Fig. 7.2) were compared to those of the conventional level shifting circuit. The maximum frequency of operation F_{max} and the average power P_{ave} of the two drivers are shown in Table 7.2 for the different level shifting techniques. F_{max} is defined as the input frequency at which the output voltage swing is reduced by about 30% from its low frequency value. These simulation results show that while, in general, F_{max} of the ECL driver is significantly increased by level shifting, F_{max} of the CML driver is not. This is because F_{max} of the CML driver is dominated by the current-switch rather than level-shifting at the inputs. However, for both drivers the novel level-shifter results in a much smaller power consumption (power is reduced by up to 60%).

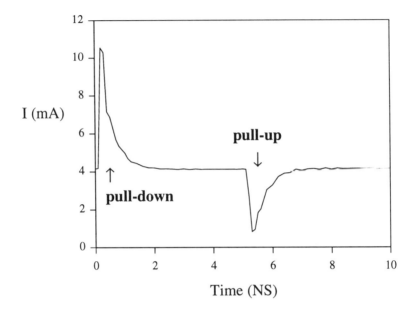

Figure 7.5 The collector current of Q_2 in fig.2(b) during the output pull-down and pull-up.

7.2.2 The DCCP Current Source

As was mentioned earlier, the CML driver is conventionally used when lower power is required. For this driver the tail current value is determined by the value of the termination resistance and the required output voltage swing. Hence, for a 50Ω termination and 0.8V swing the value of the tail current has to be set around 16 mA. This limits the speed of the driver. Also, when the output is high and the tail current is in the other branch of the current switch (assuming a single ended output) a high value of this current (and hence the power) is not essential for the proper operation of the circuit. A novel current source circuit that utilizes a dynamically-controlled biasing network to overcome these shortcomings is shown in Figure 7.4 along with the conventionally used current source circuit.

The DCCP circuit works as follows (referring to Figure 7.4(b)); The voltage Vc, which in conjunction with Rs determines the value of the tail current, is equal to the difference between the voltages at Va and Vb. So when the input

	The ECL driver [+]		The CML driver	
Level-shifter	F_{max} (GHz)	P_{ave} [*] (mW)	F_{max} (GHz)	P_{ave} [*] (mW)
Ct. 2(a)	2.35	239	1.20	206
Ct. 2(b)	2.60	114	1.25	81

[+] The tail current is 10 mA   [*] At Fmax

Table 7.2 The effects of input signals level shifting on the different performance parameters of the ECL and CML drivers.

I/P is 'High' (i.e. $= VDD - V_{BE}$), Vc is given by

$$Vc = Va - Vb, where \qquad (7.1)$$

$$Va = VDD - 2V_{BE}, and \qquad (7.2)$$

$$Vb = VDD - (1+z) \cdot V_{BE} - V_{BE}, \qquad (7.3)$$

Where z is the multiplication factor in the V_{BE} multiplier in Figure 7.4(b) (made of Q1 and the two resistors R and zR).
Hence, from the above equations we get:

$$Vc = z \cdot V_{BE} \qquad (7.4)$$

z is adjusted such that Vc is about 1.2-1.3V (i.e. z is from 1.5 to 1.8) and consequently, Rs is adjusted such that the 'High' value of the current is around 16 mA.

When I/P is 'Low', Vb is determined by the second V_{BE} multiplier in Figure 7.4(b) (made of Q2 and the two resistors R2 and xR2). So Vb becomes:

$$VDD - (1+x) \cdot V_{BE} \qquad (7.5)$$

and Vc becomes:

$$Vc = (z - x + 1) \cdot V_{BE} \qquad (7.6)$$

Where x is between 1.2 to 1.3. Also, it is assumed that the input voltage swing (V_s) is larger than $(x - 1) \cdot V_{BE}$ such that the 'Low' value of the tail current which is determined by the value of Vc in equation (7.6) above is independent of the voltage swing. This 'Low' value of the tail current constitute a trade off between power and speed. The smaller this value is, the lesser the power is, and the larger the time it takes to boost the current to the 'High' value and switch the circuit is. The capacitances Cx's are used to enhance the speed and resolve (to some degree) the above trade off. Meanwhile, they do not affect the DC characteristics.

Again as in the DAPD circuit, these capacitors are set to the highest value that would not load the driving gate (about 0.5pF in this work). The actual values of R, R1, and R2 are not important and they can be set to high values to reduce the power in the biasing network. The value of Rx represents a trade-off between speed and power. The lower Rx is, the higher both the speed and the power. So in this work Rx was set to 100 Ω which yielded a very good speed. This resulted in a 4.5 mW power consumption in the biasing network, a very negligible amount of power compared to that saved in the current-switch.

Figure 7.6 shows the simulated tail current generated by the novel circuit during the pull-up and pull-down transitions of the output of the CML driver. It shows how the tail current gets boosted during the pull-down transition and reduced during the pull-up transition.

Figure 7.7 shows the simulation results of a comparison between the power (total power and the power withdrawn from VDD) of the CML driver with the DCCP current source (including the power consumed by the biasing network) and the driver with the conventional source. As expected, at low frequencies the power difference is the largest but as the frequency increases the difference gets smaller. This is because the biasing network will have less time to bring the tail current to its 'Low' value, hence the increase in power. As the frequency increases further the difference become constant since the biasing network would just keep the tail current at a constant value. Nevertheless, Pvdd of the driver with the novel DCCP circuit would still be 40% less than that of the CML with the conventional current sourcing circuit. Also for typical operating conditions (e.g. burst modes), the DCCP circuit would save lots of idle power as compared to the conventional current source circuit.

The micrograph of the test circuit of a CML driver with the DCCP current source is shown in Figure 7.8. The current source (and its biasing network) was designed to produce a high current of about 16 mA (i.e. 0.8 V swing across the external 50 Ω load). The measured output waveform of this test

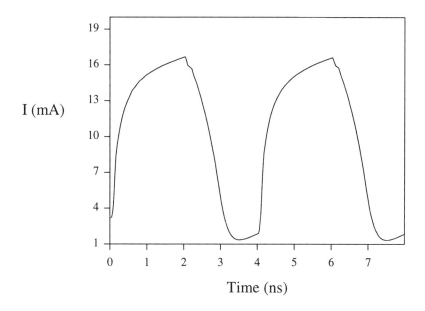

Figure 7.6 The tail current produced by the novel DCCP current source circuit during pull-up and pull-down output transitions.

circuit at 1.5 GHz is shown in Figure 7.9. This figure shows that even at this high frequency the output voltage swing is still close to 0.8V (the low-frequency value) i.e. the new current source is working perfectly at this high frequency.

The DCCP can also be used in to reduce the power of ECL drivers by reducing the power consumed in the current-switch part.

7.3 THE UNIVERSAL TRANSCEIVER (RECEIVER/DRIVER)

The novel low-voltage-swing, low-power transceiver consists of a universal receiver and a universal output driver operating with a voltage supply of 3.3V. The receiver can read signals with termination voltages ranging from 1.5V to 5V and the driver can drive an external 25 Ω terminated to a termination voltage (V) ranging from 2V to 5V. Both circuits do not require the use of any

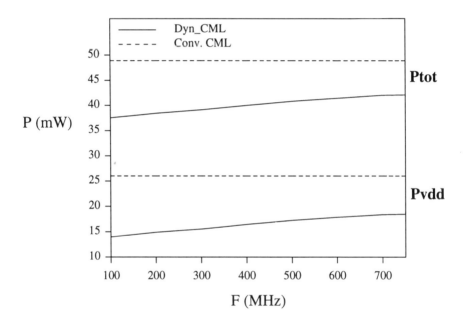

Figure 7.7 The total power (Ptot) and the power withdrawn from VDD (Pvdd) vs frequency for the CML driver with the dynamic current source (Dyn-CML) and the conventional CML driver.

Figure 7.8 The micrograph of the CML driver with the new DCCP current source.

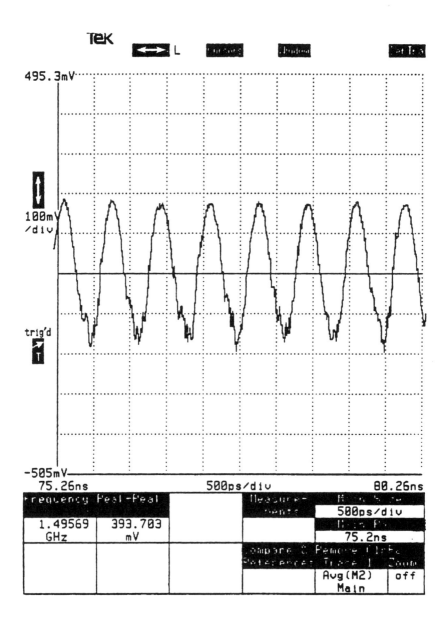

Figure 7.9 The measured output waveform of a CML driver with the new DCCP current source at 1.5 GHz. The output is 50Ω terminated to 3.3V (6db attenuation at the input of the sampling scope).

external reference voltages. Also, the only restriction imposed on the signals is that they are not level-shifted with respect to the termination voltage VT. Hence assuming the signal swing (V_s) to be between 0.8V to 1.0V, all signals should be from VT to VT - V_s.

The circuit description, operation, and performance of the receiver and the driver are presented below.

7.3.1 The Universal Receiver

The novel receiver circuit consists of three sub-circuits; the universal input buffer (Figure 7.10), the Vref generator (Figure 7.11(a)), and the load control circuit (Figure 7.11(b)). The description and the operation of each of these subcircuits are presented below. This would be followed by the results of the performance evaluation of the whole receiver.

7.3.1.1 The Universal Input Buffer (UIB)

The UIB is basically an emitter-coupled BJT pair followed by a source-follower stage. Four PMOS devices M1-4 (Figure 7.10) are used as loads for the emitter-coupled BJT's. These load devices are controlled by the two biasing voltages V1 and V2 which are generated by the load control circuit (Figure 7.11(b)). The two PMOS devices M5 and M6 ensure that the N-wells of all the PMOS devices M1-6 are connected to the highest voltage among VT and VDD (3.3V). The value of the reference voltage Vref in Figure 7.10 is kept at VT - 0.45 V by the Vref generator circuit. A source-follower stage was used instead of an emitter-follower to avoid saturating the emitter-followers' BJTs when VT is larger than VDD. This also ensures that the UIB differential output signal will not saturate the driven gates (which use a 3.3V supply). However, the speed is slightly reduced.

The BJT emitter-coupled pair will steer the tail current I_{SS} between the two branches of the UIB depending on the input signal level. For a VT less than VDD, M1 and M4 will be off, and M2 and M3 will be on and act as loads for the differential pair in the current-switch. Hence, the output signal of the differential pair (before the source-follower) will be from VDD to $VDD - V_s$. Also, the N-wells of the PMOS devices M1-6 will be connected to VDD via M5. Similarly, when VT is greater than VDD (3.3V), M2 and M3 will be off, M1 and M4 will be on, the N-wells of M1-6 will be connected to VT via M6, and the differential pair output swing will be from VT to $VT - V_s$.

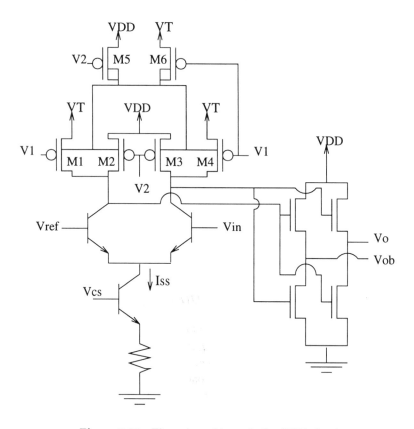

Figure 7.10 The universal input buffer (UIB) circuit.

The sizes of M5 and M6 can be set to a minimum. The sizes of M1-4 are adjusted depending on I_{ss} to give a swing of about 0.5V. V_{GS} of the load devices is constant for VT's less than 3.3V. When VT exceed 3.3V, V_{GS} of the turned-on load devices (M1 and M4) increases with VT and hence the voltage swing of the UIB decreases. However, because these device are in the linear region to start with (and remain so), the change in the outputs voltage swing is very small (less than 15and the output differential signals still maintain very good noise margins. Figure 7.12 shows the simulation results for the differential outputs of the UIB for two termination voltages (2V and 5V) at 1GHz and a Fan_{OUT} of one ECL gate at the output. The value of the tail current of the UIB used for these simulations was 1 mA. This figure shows how the magnitude of the output swing changes slightly as VT increase beyond VDD. It also shows

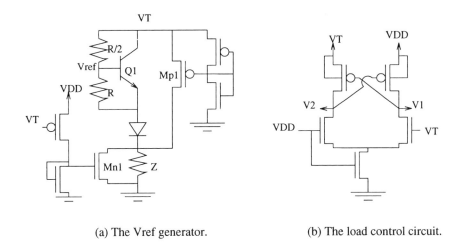

(a) The Vref generator. (b) The load control circuit.

Figure 7.11 The two reference circuits in the universal receiver.

how the outputs levels shift with VT. Due to the differential nature of the output, this shift with VT is not important unless it saturate the BJT devices in the driven ECL gate. As Figure 7.12 shows, for the maximum VT (5V), the output levels are still acceptable and would not saturate the input devices of any ECL or CML gate. Also, for VT's less than 3.3V, the output levels will still be the same as those in Figure 7.12 (the bottom trace). These levels are high enough not to cause the saturation of the current source of driven ECL or CML gates.

The maximum frequency of operation Fmax (defined here as the frequency at which the output swing of the UIB becomes less than 250 mV) of the UIB versus the value of I_{ss} is shown in Figure 7.13. For these simulations the sizes of the load PMOS devices M1-4 were proportionally increased with the tail current to keep the output voltage swing of the UIB constant. This figure shows that there is an optimum value of the tail current that yield a maximum Fmax.

7.3.1.2 The Vref Generator

The UIB requires a reference voltage, Vref, that is alway about 0.45V below the termination voltage (i.e. Vref = VT - 0.45V) for VT's ranging from 1.5V to 5V. Also This reference voltage should be stable over a large temperature

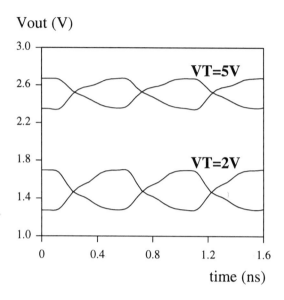

Figure 7.12 Simulation output waveforms of the UIB for two termination voltages (2V and 5V) at 1GHz and an ECL gate at the output.

range. Standard referencing circuits such as Band Gap reference circuits can not be designed to meet the first criteria.

The Vref generator circuit of Figure 7.11(a) was designed to meet the above two requirements. Vref, which is about $\frac{1}{2}V_{BE}$ below VT, is generated by the V_{BE} multiplier (made of Q1, R, and R/2). Mn1 is only on for VT's less than 2.5V and is used to compensate the decrease in Q1 current. The Mp1 and the diode is used for temperature compensation. Mp1 is biased in such a way that its drain current will increase linearly with VT. This will keep the current through Q1 constant and hence Vref will always remain constant with respect to VT. Also, as the temperature increases, the drain current of Mp1 decreases and hence the emitter current of Q1 will increase and compensate the decrease in V_{BE} that occur as the temperature increases (which is about 2.5 mV/C^o). The reverse happens when the temperature decreases.

The layout of the fabricated Vref circuit is shown in Figure 7.14. This figure shows the compact area of the Vref generator circuit.

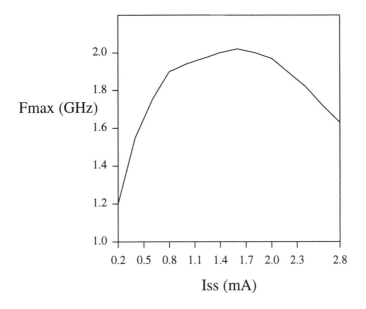

Figure 7.13 The maximum frequency of operation (Fmax) of the UIB vs the tail current (I_{ss}).

Figure 7.15 shows the measurement results for the output of the Vref generator versus temperature for several termination voltages. This figure shows the stability of Vref over a wide range of temperatures and termination voltages. The maximum change in Vref over the whole temperature range for all values of termination voltages was about 51 mV (about 5% of the voltage swing).

The Vref generator circuit was designed such that the difference between Vref and VT is 450 mV for all values of VT. Figure 7.16 shows the measurement results for the difference (VT - Vref) versus VT at room temperature (25 C^o). Two sets of data are shown in that figure. These two sets of data were measured from two, randomly selected, different dies that came from two different wafers that were fabricated in two separate runs. This figure not only shows how accurate the generated Vref is for all of the termination voltage range, but also how robust and process insensitive the Vref generator circuit is. The measurements from the two wafers are very consistent, and for most values of VT, the maximum deviation of Vref from the required value is less than 20 mV. The largest deviation in Vref occurred at VT's between 2V and 2.5V.

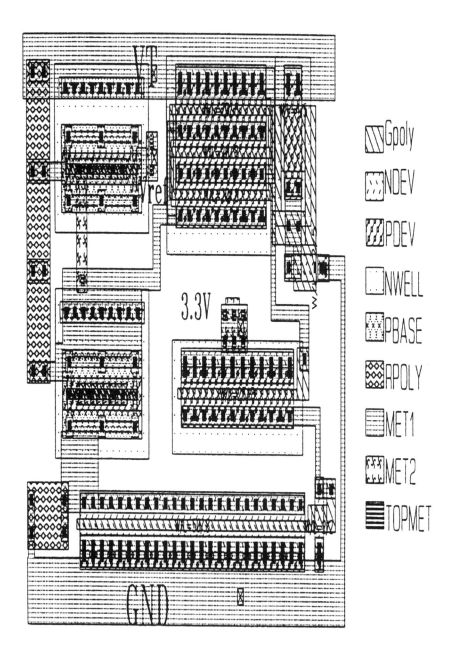

Figure 7.14 The layout of the Vref generator circuit.

Inter-Chip Low-Voltage-Swing Transceivers 173

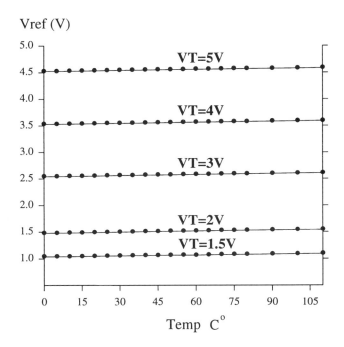

Figure 7.15 Measurement results showing the output of the Vref generator vs temperature for different termination voltages.

However, the deviation is still less than 50 mV and consistent between the two wafers.

7.3.1.3 *The Load Control Circuit*

This circuit provide the control biasing voltages for the PMOS load devices in the UIB . It is a simple source-coupled NMOS pair with cross-coupled PMOS loads. The way the input and load MOS devices are connected and sized ensures that when VT is less than VDD V1 will be 'High' and close to VDD (VDD - V1 would be less than the PMOS threshold voltage V_{tp}). At the same time V2 will be 'Low'. When VT is greater than VDD V2 will be 'High' and close to VT (VT - V2 < V_{tp}), and V1 will be 'Low'. So for VT < VDD M1 and M4 in the UIB are off, and M2 and M3 are on. M5 will also be on and the N wells will be connected to VDD. If VT exceeds VDD M2, M3, and M5 will be turned

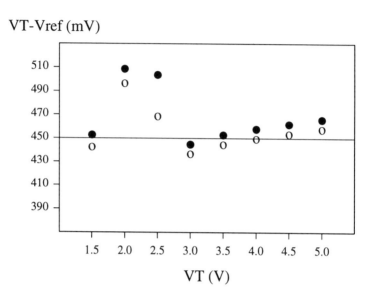

Figure 7.16 Measurement results showing VT - Vref vs VT at room temperature. The two sets of data are from two different wafers fabricated in separate runs.

off and M1, M4 ,and M6 will be turned on. The N wells will be connected to VT. As VT increases further V2 keeps increasing such that M2, M3, and M5 remain off. This in turn ensures the correct operation of the UIB. No current will be flowing between VT and VDD under any circumstances.

The operation of the load control circuit is verified in Figure 7.17 where the outputs V1 and V2 are shown versus VT. The load control circuit was found to be very stable over the temperature range. The maximum change in V1 and V2 over the temperature range for any value of VT was found to be less than 10 mV.

7.3.1.4 *The Receiver Performance*

The universal transceiver was fabricated using NT's $0.8\mu m$ BiCMOS process. The layout of the UIB and the load control circuit is shown in Figure 7.18. The micrograph of the receiver circuit test structure is shown in Figure 7.19. The

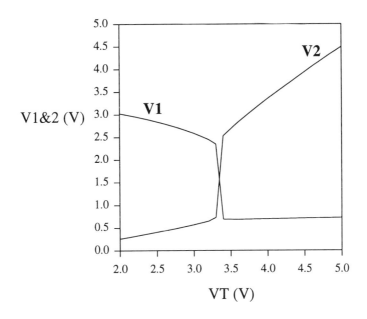

Figure 7.17 The outputs of the load control circuit (V1 and V2) vs the termination voltage VT.

test structure consists of a universal receiver driving an ECL driver, which in turn drives the outputs (the differential outputs are 50Ω off-chip terminated to ground). The tail current I_{ss} of the UIB in the receiver was set to 1 mA.

The measured differential output waveforms of the ECL driver (being driven by the universal receiver) at a frequency of 1 GHz for three different input termination voltages (5V, 4V, and 3V) are shown in Figures 7.20-7.22. Figure 7.23 shows the same outputs at 1.5 GHz and a termination voltage of 2V. These figures verify the operation of the receiver for all the termination voltage range at high frequencies and its ability to switch the ECL driver. The total power of the receiver with the 2V termination voltage was 12.5 mW.

7.3.2 The Universal Output Drivers

Two versions of the universal output driver were developed; the first version (UOD1) is shown in Figure 7.24, and the second version (UOD2) is shown in

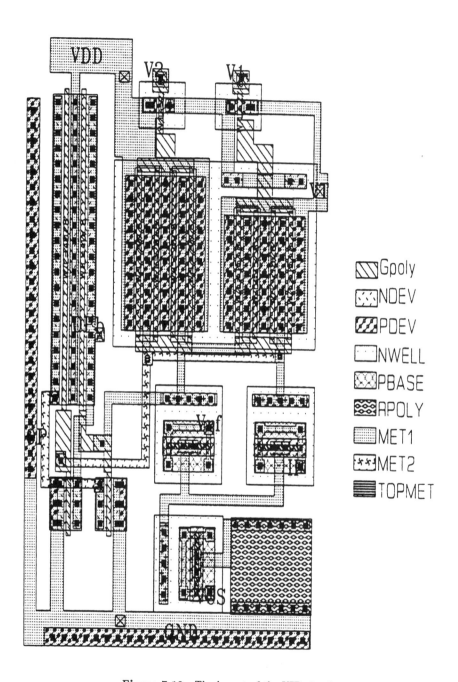

Figure 7.18 The layout of the UIB circuit.

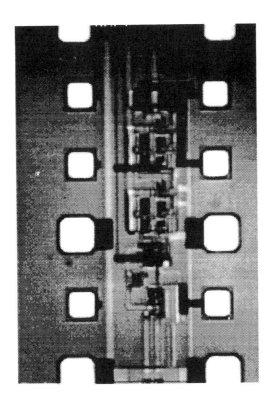

Figure 7.19 The micrograph of the universal transceiver test circuit.

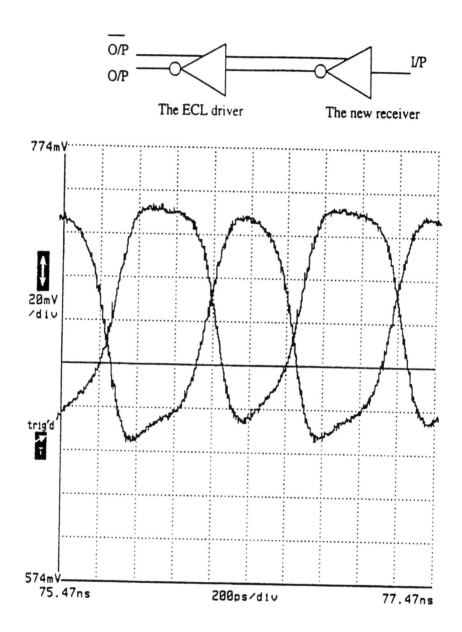

Figure 7.20 The measured output waveforms of an ECL driver being driven by the new receiver at 1 GHz. The input signal to the receiver is terminated to 5V (attenuation at the sampling scope inputs is 20db).

Inter-Chip Low-Voltage-Swing Transceivers

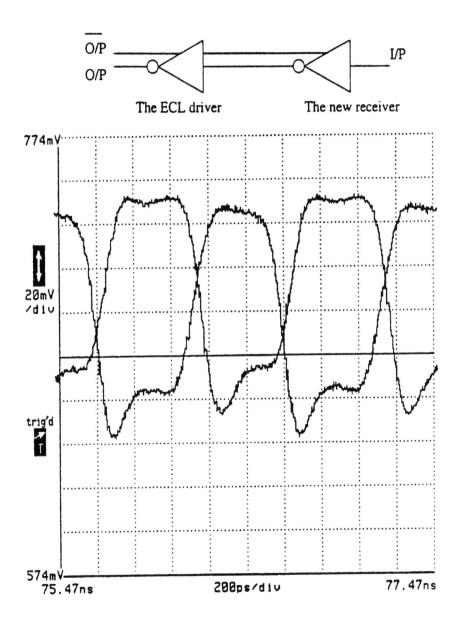

Figure 7.21 The measured output waveforms of the new receiver test structure at 1 GHz with 4V input signal termination (attenuation at the sampling scope inputs is 20db).

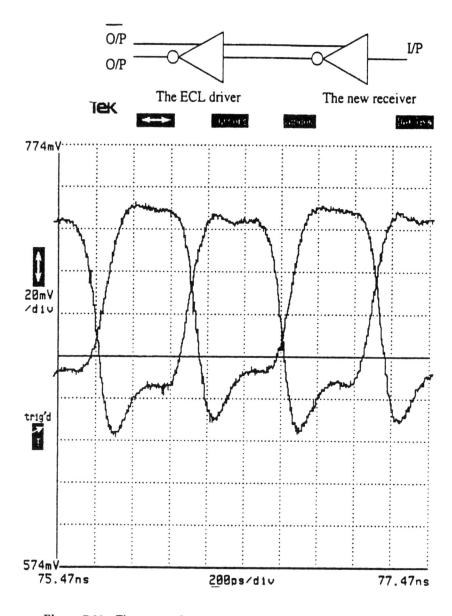

Figure 7.22 The measured output waveforms of the new receiver test structure at 1 GHz with 3V input signal termination (attenuation at the sampling scope inputs is 20db).

Inter-Chip Low-Voltage-Swing Transceivers 181

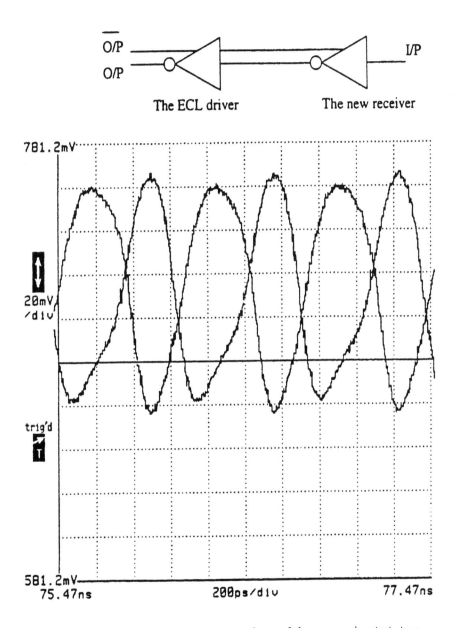

Figure 7.23 The measured output waveforms of the new receiver test structure at 1.5GHz with 2V input signal termination (20 db attenuation at the sampling scope inputs).

Figure 7.25. For the design of both drivers, double termination (25Ω load to VT) was assumed. The circuit description, operation, and design of each driver are provided below, followed by the performance evaluation of both drivers.

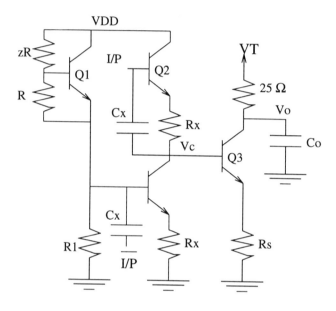

Figure 7.24 The first version of the universal output driver (UOD1).

7.3.2.1 The UOD1

The UOD1 circuit is very similar to the DCCP circuit except that it was slightly modified such that the 'Low' value of the tail current becomes zero. To achieve this Q2, R2, and xR2 (in Figure 7.4(b)) were removed, as Figure 7.24 shows, and the following condition was imposed on z:

$$1 < z \leq 1 + \frac{V_s}{V_{BE}} \qquad (7.7)$$

Again, z is the multiplication factor of the V_{BE} multiplier made of Q1, R and zR. The above condition will ensure that the 'Low' value of Q3 current (the tail current) is zero (i.e. when I/P is low Vc would be less than or equal to V_{BE}). Hence, when I/P is high :-

$$Vc_{(high)} = z \cdot V_{BE} \leq V_{BE} + V_s \qquad (7.8)$$

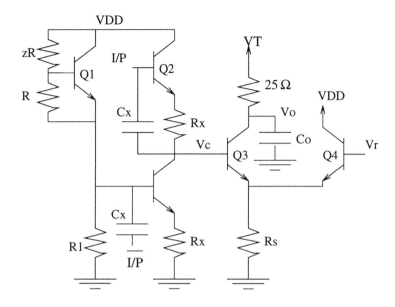

Figure 7.25 The second version of the universal output driver (UOD2).

so for this driver a rough value of V_s should be known in advance so that it can be designed properly. The 'High' value of the tail current, however, is still independent of the exact value of V_s. Rs is adjusted in conjunction with the high value of Vc to give a 'High' value of the tail current of about 32 mA.

This circuit has two shortcomings; 1) The large BJT Q3 is turned-on/off solely through the base, hence the base resistance R_B makes the turn-on/off slower than combined base/emitter turn-on/off (such as in CML/ECL circuits), and 2) The 'Low' value of the tail current might not be exactly zero although it is designed to be so (it may reach about a few hundreds of μA's) due to the tolerance in the resistors in the biasing network. This would slightly reduce the noise margins by reducing the value of V_oH (the high output voltage).

7.3.2.2 The UOD2

To overcome the two shortcomings of the UOD1 mentioned above the UOD1 circuit was slightly modified to produce the second version of the universal output driver, UOD2, shown in Figure 7.25. Another BJT, Q4 (which is much smaller than Q3), was added with its base connected to a voltage reference, Vr. For this version the restriction on z is relaxed since the 'Low' value of Vc (Vc_{Low}) does not have to be less than V_{BE}. The 'High' value of Vc is still equal to $z \cdot V_{BE}$. When the input I/P goes low the tail current is steered away from Q3 to Q4. The 'Low' value of the tail current, however, is much smaller than the 'High' value due to the following reasons; 1) Vr is set to a value less than the 'High' value of Vc, and 2) V_{BE} of Q4 is larger than that of Q3 (due to the size difference). Hence, the voltage drop across Rs would be much smaller when I/P is low. When I/P becomes high the 'Low' tail current is steered away from Q4 to Q3 and helps turning Q3 on. Also, since Vc_{Low} can be made greater than that of the UOD1 without affecting V_{oH} and the noise margins, the voltage swing of Vc could be made smaller, and hence the circuit would operate faster. Again the value of the 'Low' tail current represent a trade-off between power and speed.

The disadvantages of the UOD2 as compared to the UOD1 are the slight increase in power consumption area.

7.3.2.3 Performance Evaluation of The UODs

The simulation results showed that the UOD2 speed is superior to that of the UOD1. Figure 7.26 shows the maximum frequency of operation, Fmax (defined as the input frequency at which the driver's output swing is reduced to 70% of its low-frequency value) versus VT. This figure shows that the UOD2 has a higher Fmax than the UOD1 over the whole VT range. It also shows that for both drivers Fmax decreases as VT decreases due to the increase of the effective collector capacitance and the saturation effects of Q3.

The fabricated test structures for both drivers consist of two ECL pre-drivers each driving one of the UODs, Figure 7.27. The two ECL pre-drivers driving the UODs are identical and are being driven by the same external source signals. These signals, however, were not connected at the same manner to both drivers which resulted in the outputs of the ECL drivers (and hence the UODs) being complementary to each other. This made it difficult to measure the delay difference between the two UODs. Also, as mentioned before, the UODs outputs were externally terminated to VT (the termination voltage) via a 25Ω resistor.

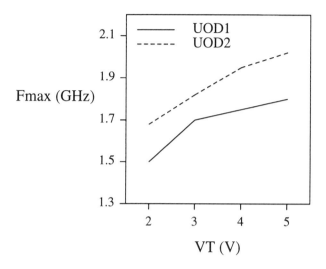

Figure 7.26 The maximum frequency of operation of the two UODs vs VT.

Figure 7.28 shows the outputs of both drivers with a termination voltage of 5V and at a low frequency (25 MHz). This figure shows the sharper edges of the UOD2 output. It also shows that the UOD2 has a slight overshoot and undershoot in its output.

Figure 7.29 shows the outputs of both drivers with the termination voltage being 2V at the same low frequency. This figure shows how the UOD2 suffers under very low termination voltage conditions. Its output voltage swing deteriorates significantly and its delay increases (especially, the rising edge) due to the saturation (of Q3) effects. The output of the UOD1 is hardly affected at all by the lowering of the termination voltage from 5V to 2V as evident from Figures 7.28 and 7.29. This is an added advantage for the UOD1 over the UOD2.

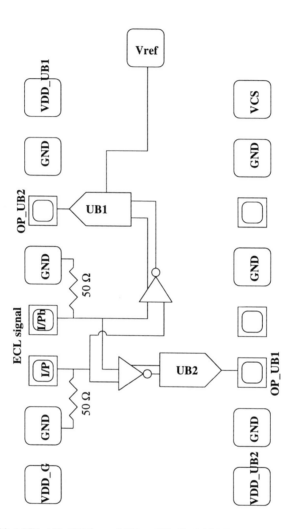

+ VCS is 1.25V, ALL VDD's are 3.3V, and Vref is 1.15-1.2 V.

* **UB1_OP and UB2_op are 25 Ohms terminated to VT, VT ranges from 5V to 2V.**

Figure 7.27 Schematic of the test structures of the two UODs.

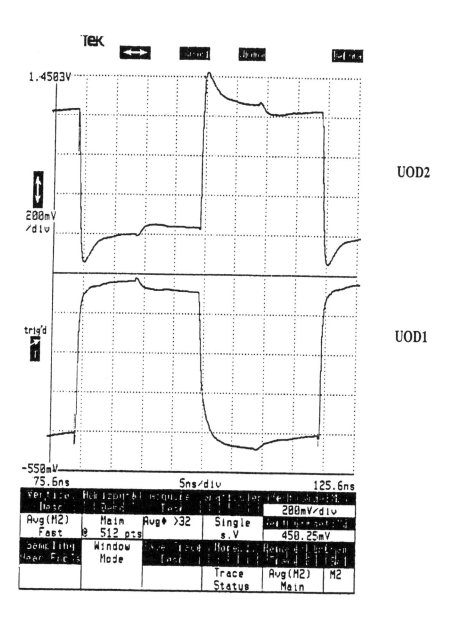

Figure 7.28 The measured output waveforms of the two UODs with a 5V termination voltage and a 25 MHz input frequency.

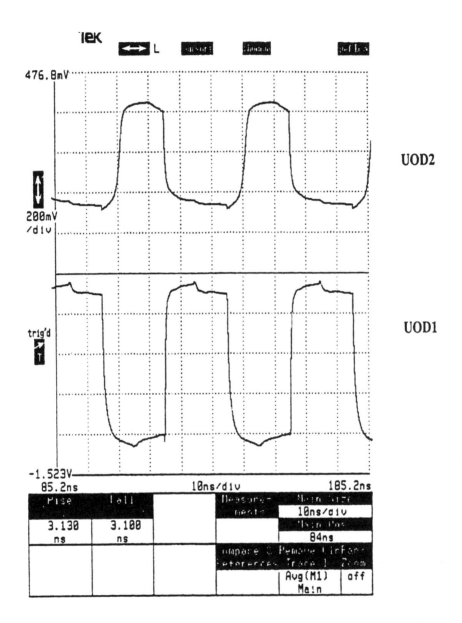

Figure 7.29 The measured output waveforms of the two UODs with a 2V termination voltage and a 25 MHz input frequency.

Figure 7.30 shows the output of both drivers at 500 MHz and a termination voltage of 5V. These waveforms were measured from a die different than the one that was used to produce the results in Figures 7.28 and 7.29. Again the second die was from a different wafer fabricated in a different run than the first one. This figure shows the slight reduction in both UODs output swings from the results obtained from the first die (Figures 7.28 and 7.29). The reduction in the UOD1 output swing, however, is significantly larger than that of the UOD2 as was predicted before. The UOD1 output swing is more sensitive to process variation than the UOD2. However, due to the fact that only two samples were tested, no solid conclusion can be made as how sensitive to process variation the UOD1 actually is. The UOD2, on the other hand is showing significant consistency. The on-chip powers at that frequency were 75 mW and 80 mW for the UOD1 and the UOD2, respectively. The power of a CML driver at the same termination voltage would be about 125 mW (not including the power of biasing/referencing circuits). An ECL or pseudo ECL driver would have an even greater power at the same conditions. For a 2V termination the UOD1 and the UOD2 powers are 25 mW and 29 mW, respectively.

The measured output waveform of UOD1 at 1 GHz (using the first die) is shown in Figure 7.31. The termination voltage used was 5V. As this figure shows, the output voltage swing at this high frequency is still about 0.8V (the same as the low-frequency swing).

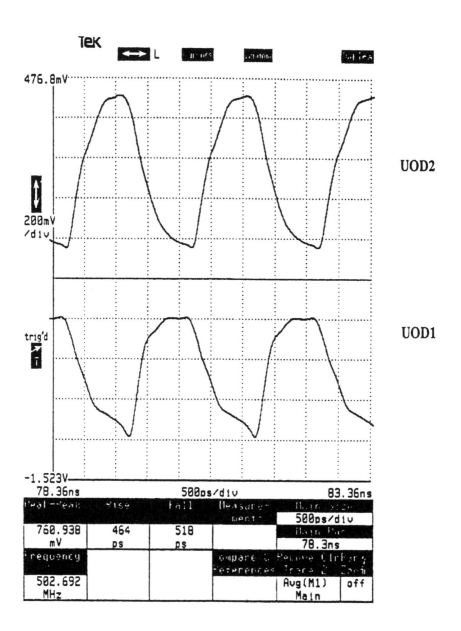

Figure 7.30 The measured output waveforms of the two UODs at 500 MHz and a VT of 5V. The above results were produced from a different die than the one used for Figures 7.28 and 7.29 results.

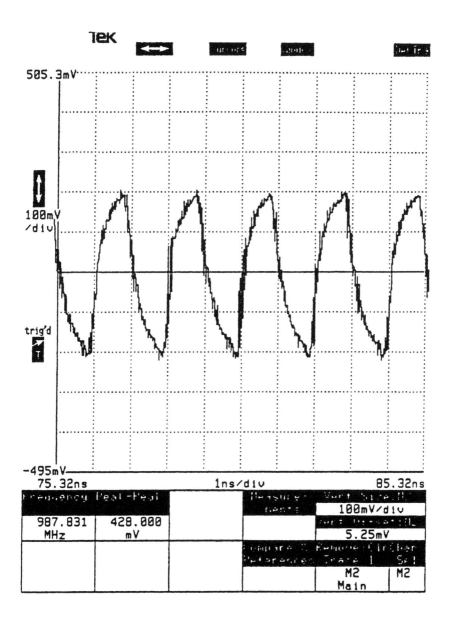

Figure 7.31 The measured output waveform of the UOD1 at 1 GHz and 5V termination (6db attenuation at the sampling scope input).

7.4 CHAPTER SUMMARY

Novel circuit techniques that take advantage of the existing BiCMOS technologies and enhance inter-chips communications were developed. Two circuit techniques that enhance the speed of ECL/CML drivers while reducing their power consumption were presented. These are; the dynamic active pull-down level shifting/buffering circuit, and the dynamically-controlled charge-pumped current source circuit. It was shown that the DAPD level-shifter circuit can achieve very high speeds at a much lower power than the conventional emitter-follower level-shifting circuit. Although the CML driver did not benefit from the DAPD as much as the ECL driver in terms of speed, the power savings are still enormous. The new DCCP current sourcing technique would benefit both types of drivers in terms of power saving. Also, a novel low-voltage-swing transceiver with low on-chip-power was developed. It was demonstrated that the novel transceiver can operate with termination voltages ranging from 2 to 5 volts without the need for an off-chip reference voltage. All the needed referencing and biasing voltages are generated on-chip. The good performance of the biasing circuits was demonstrated over a wide range of temperatures and termination voltages. The biasing circuits also proved to be very robust and process insensitive. Two versions of the driver were developed. The first version works better for low termination voltages but is more sensitive to process parameters. Both drivers have lower on-chip-power than ECL/CML or pseudo ECL drivers.

Experimental results from measurements performed on test chips of the new circuits were provided throughout the chapter.

This work demonstrated the capabilities of the BiCMOS technology to implement diversified, high performance, smart I/O's and voltage referencing circuits.

REFERENCES

[1] R. Hadaway, *et al.*, "BiCMOS Technology for Telecommunications," *IEEE BCTM Proc.*, pp. 159-166, 1993.

[2] John H. Quigley, *et al.*, "Current Mode Transceiver Logic (CMTL) for Reduced Swing CMOS, Chip to Chip Communication," *IEEE Int. ASIC Conf. and Exhibit Proc.*, pp. 452-455, 1993.

[3] M. Pedersen, and P. Metz, *et al.*, "A CMOS to 100K ECL Interface Circuit," *ISSCC Tech. Dig.*, pp. 226-227, 1989.

[4] T. J. Gabara and S. Knauer, "Digital Transistor Sizing Techniques Applied to 100K ECL CMOS Output Buffers," *IEEE Int. ASIC Conf. and Exhibit Proc.*, pp. 456-459, 1993.

[5] Bill Gunning, *et al.*, "A CMOS Low-Voltage-Swing Transmission-Line Transceiver," *ISSCC Tech. Dig.*, pp. 58-59, 1992.

[6] C. T. Chuang, and D. D. Tang, "High-Speed Low-Power AC-Coupled Complementary Push-Pull ECL Circuit," *IEEE J. Solid-State Circuits*, Vol. 27, pp. 660-663, 1992.

[7] C. T. Chuang, *et al.*, "High-Speed Low-Power ECL Circuit With AC-Coupled Self-Biased Dynamic Current Source and Active-Pull-Down Emitter-Follower Stage," *IEEE J. Solid-State Circuits*, Vol. 27, pp. 1207-1210, 1992.

[8] W. Wilhelm, and P. Weger, "2V Low-Power Bipolar Logic," *ISSCC Tech. Dig.*, pp. 94-95, 1993.

[9] H. Shin, "A Self-Biased Feedback-Controlled Pull-Down Emitter Follower for High-Speed Low-Power Bipolar Logic Circuits," *IEEE J. Solid-State Circuits*, Vol. 29, pp. 523-528, 1994.

Index

Adders
 32-Bit Macro, 7
 64-bit Adder, 40
 Carry Select (CS), 7
 Carry-Save Adder Array, 40
 Circuit Styles, 17
 Comparison, 17
 Conditional Sum Addition
  (CSA), 7, 33
  Architecture, 9
  Average Power, 17
  CPL-Like Implementation, 12
  Critical Path, 9
  Layout, 24
  Measurements, 26
  Minimum Power, 16
  MUX Implementation, 14
  Optimization, 14
  Power Estimation, 16
  Propagation Delays, 9
  Transmission Gate (TG)
   Implementation, 12
 Energy, 17
 Power Dissipation, 20
 Supply Sensitivity, 20
BiCMOS Circuits
 Low-voltage-swing, 154
BiCMOS Drivers, 125
 AND Gates, 142
 Circuit Simulations, 132
 D flip-flop, 144
 Device Simulations, 133
 Full Swing, 125
 Merged Devices, 130

 Latch-up, 134
 Partial Swing, 125
 Positive Feedback, 127
  Concept of Operation, 128
  Design, 148
Digital Signal Processors (DSP), 7
Digital Signal Processing (DSP), 31
Discrete Cosine Transform, 31
ECL/CML Circuits, 154
 Dynamic Active-pull-down
  (DAPD), 155
 Dynamically-controlled
  Charge-pumped (DCCP, 155
 Inputs Buffering, 157
Low-Power
 Adders, 7
Multipliers, 31
 6-Bit Multiplier, 71
  Floorplan, 73
  Layout, 74
 Baugh-Wooley Multiplier, 33
 Braun's Array Multiplier, 33
 Capacitance Estimation, 45
 Comparison, 42
 Compressors, 39–40
 Full Adders (FA), 31
 Modified Booth Algorithm, 34, 38
  Add Cell, 70
  Booth Encoder, 40, 64
  Booth Encoders, 43
  Floorplan, 38
  With Wallace Tree, 40

Modified Booth Multiplier, 43
Modified Radix-2 Algorithm, 35
Multiplier Cell, 47
  Full Adder (FA), 47
  Implementations, 62
  Multiplexer (MUX), 58
  Optimization, 54
  Performance, 62
  Partial Products, 35–36, 39–40
  Sign Extension, 38, 43
  Wallace Tree, 39–40
Register File, 83
  32x32-bit Layout, 95
  32x32-bit Simulations, 91
  Address Register, 89
  Decoder, 89
  Floorplan, 84
  Memory Cell, 86
  Multi-Ports, 84
  Read Circuitry, 89
    Sense Latch, 89
  Write Circuitry, 86
SRAMs
  BiCMOS ECL, 101
  BiCMOS ECL
    Block Size, 109
    Front-End, 101,,,
    Latched Sense-Amplifier, 114
    Power Optimization, 109
    Wired-OR Pre-decoders, 106
    Wired-OR &
      Level-Translation, 107
    Word-Line Decoder and Driver
      (WLDD), 111
  BiCMOS, 99
  Bipolar, 99
  CMOS, 99
  Column Sensing, 113
  Localized Self-Resetting, 111
  power limitations, 99
  Self-Resetting Blocks, 111
  Sense-Amplifiers, 119

Word-line Active-Time
  Reduction, 110
Transceivers, 153
Universal Transceiver, 163
  Input Buffer, 167
    Swing, 168
  Load Control, 167, 173
  Output Drivers, 175
  Receiver, 167, 174
  Signal Swing, 167
  Source-follower, 167
  Voltage Reference, 167

Advanced Low-Power Digital Circuit Techniques

covers contemporary problems in low-power digital VLSI circuit design. Combining different system perspectives such that of microprocessors, data transceivers, and SRAMs it offers the readers practical circuit solutions for power reduction of adverse systems. Different techniques for power reduction on both the circuit level and the block architectural level are presented. The efficient use of the available technologies such as sub-micron CMOS and BiCMOS in circuit design is superbly demonstrated throughout this book. The book provides a wide range of practical circuit designs that can serve as examples for low-power circuit designers for different VLSI subsystems. These design examples include 32-bit fast adders, parallel multipliers, high speed register files, on-chip drivers, BiCMOS SRAMs' building blocks, and high speed chip-to-chip communication circuits. Many experimental results from measurements on fabricated circuits are included to validate the operation and performance of these circuits. Written in plain and clear language, this book can benefit design engineers, post graduate, and senior undergraduate university students.

About the Authors

M. S. Elrabaa earned his Ph.D. degree from the Department of Electrical and Computer Engineering, University of Waterloo, Waterloo, Ontario, Canada. He authored and co-authored several papers on digital BiCMOS design methodologies and novel BiCMOS circuits for different applications such as logic, SRAMs, and Telecommunications. He is currently a senior component design engineer with Intel corporation. His current interest is in low-power Phase-Locked-Loops design.

I. S. Abu-Khater earned his Ph.D. degree from the Department of Electrical and Computer Engineering, University of Waterloo. He authored and co-authored several papers on digital CMOS and BiCMOS circuit design for low-voltage low-power VLSI applications with emphasis on microprocessor's building blocks. He is currently a senior component design engineer with Intel corporation. His current interest is in on-chip caches for high-speed microprocessors.

M. I. Elmasry is Professor of Electrical and Computer Engineering, at the University of Waterloo, Waterloo, Ontario, Canada and the founding director of the VLSI Research Group. He has published 13 books, over 250 research papers and has been a consultant to many R&D labs in the area of VLSI design of digital circuits and systems. He is a Fellow of the IEEE for his contributions to that area.